International Conference on Biomaterials, Nanomaterials and Composite Materials 2016 (CBNCM 2016)

MATEC Web of Conferences Volume 88 (2016)

Chengdu, China
4 - 6 November 2016

Editor:

L. Zhang

ISBN: 978-1-5108-3364-7

Printed from e-media with permission by:

Curran Associates, Inc.
57 Morehouse Lane
Red Hook, NY 12571

Some format issues inherent in the e-media version may also appear in this print version.

Printed by Curran Associates, Inc. (2017)

For additional information, please contact EDP Sciences – Web of Conferences
at the address below.

EDP Sciences – Web of Conferences
17, Avenue du Hoggar
Parc d'Activité de Courtabœuf
BP 112
F-91944 Les Ulis Cedex A
France

Phone: +33 (0) 1 69 18 75 75
Fax: +33 (0) 1 69 28 84 91

contact-edps@webofconferences.org

Additional copies of this publication are available from:

Curran Associates, Inc.
57 Morehouse Lane
Red Hook, NY 12571 USA
Phone: 845-758-0400
Fax: 845-758-2633
Email: curran@proceedings.com
Web: www.proceedings.com

TABLE OF CONTENTS

ISSN (electronic edition): 2261-236X

COMMITTEES

Conference Committee Chairs

Prof. Stephen M. Hsu, The George Washington University, USA

Prof. Linmao Qian, Southwest Jiaotong University, China

Program Committee Chairs

Prof. Liangchi Zhang, University of New South Wales, Australia

Prof. Stojan Djokic, University of Alberta, Canada

International Technical Committee

Prof. CHEN Jingsheng, National University of Singapore, Singapore

Prof. Kien Wen Sun, National Chiao Tung University, Taiwan

Dr. Sushanta Kumar Panigrahi, Indian Institute of Technology Madras, India

Dr. Jun Yang, University of Western Ontario, Canada

Dr. SU Xiaodi, A-Star, Singapore

Prof. Li Chang, The University of Sydney, Australia

Prof. Mohd Sapuan b. Salit , Universiti Putra Malaysia, Malaysia

Prof. Laoui Tahar, King Fahd University of Petroleum & Minerals, Saudi Arabia

Dr. Azade Yelten, Istanbul University , Turkey

Statement of Peer review

In submitting conference proceedings to MATEC Web of Conferences, I certify to the Publisher that I adhere to the **Policy on Publishing Integrity** of the journal in order to safeguard good scientific practice in publishing.

1. All articles have been subjected to peer review administered by the proceedings editors.
2. Reviews have been conducted by expert referees, who have been requested to provide unbiased and constructive comments aimed, whenever possible, at improving the work.
3. Proceedings editors have taken all reasonable steps to ensure the quality of the materials they publish and their decision to accept or reject a paper for publication has been based only on the merits of the work and the relevance to the journal.

Title, date and place of the conference

2016 International Conference on Biomaterials, Nanomaterials and Composite Materials (CBNCM 2016)
4-6 November 2016
Chengdu, China

Proceedings editor(s):

Liangchi Zhang

Date and editor's signature 23 November 2016

PREFACE

CBNCM 2016 was held in Chengdu during November 4-6, 2016. CBNCM 2016 was organized by Hong Kong Society of Mechanical Engineers (HKSME) and Technical Sponsored by Southwest Jiaotong University, China. The conference provides a useful and wide platform both for display the latest research and for exchange of research results and thoughts in broad Biomaterials, Nanomaterials and Composite Materials. The participants of the conference were from almost every part of the world, with background of either academia or industry, even well-known enterprise. The success and prosperity of the conference is reflected high level of the papers received.

The proceedings are a compilation of the accepted papers and represent an interesting outcome of the conference. This book covers 3 chapters: Processing and Manufacturing Technologies of materials; Nano-materials and Composite materials; Physical properties of materials.

We would like to acknowledge all of those who supported CBNCM 2016. Each individual and institutional help were very important for the success of this conference. Especially we would like to thank the organizing committee for their valuable advices in the organization and helpful peer review of the papers.

We sincerely hope that CBNCM 2016 will be a forum for excellent discussions that will put forward new ideas and promote collaborative researches. We are sure that the proceedings will serve as an important research source of references and the knowledge, which will lead to not only scientific and engineering progress but also other new products and processes.

Prof. Liangchi Zhang, University of New South Wales, Australia

Effect of UV Radiation by Projectors on 3D Printing

Iaroslav Kovalenko[1,a] and Maryna Garan[1]

[1]*Department of Manufacturing Systems and Automation, Technical University of Liberec, 46117 Liberec, Studentska 1402/2, Czech Republic*

Abstract. Polymers that solidify under light radiation are commonly used in digital light processing (DLP) 3D printing. A wide range of photopolymers use photoinitiators that react to radiation in range of ultraviolet (UV) wavelength. In the present study we provided measurement of radiant fluence in the UV wavelength range from 280 nm to 400 nm for two data projectors and compared effect of radiation on quality of 3D printing. One projector is commonly used DLP projector with high energy lamp. Second one is an industrial projector, in which RGB light emitting diodes (LEDs) are replaced by UV LEDs with wattage at the level of 3.6 % of the first one. Achieved data confirmed uneven distribution of radiant energy on illuminated area. These results validate, that undesired heating light causes internal stress inside built models that causes defects in final products.

1 Introduction

Additive manufacturing (AM) technology becomes more spread not only in prototyping but also in a small quantity production. For example, NASA uses it in space science for customizing tools, development of multifunction devices for life support and radiation shielding [1]. It's possible, because AM offers alternative approaches than traditional manufacturing.

This method of production starts its history in 1986 from process called Stereolithography (SL) [2]. It used an UV laser for solidifying of photopolymer resins. Laser is placed above the bath with build material. Laser beam moves on the surface and solidifies resin point by point. After finishing the layer, platform with solid material moves down and laser starts its next path. The main disadvantage of this method is necessity to have a full tank of photopolymer resin which is not cost-effective. Sony/D-MEC company had started producing of commercial versions of SL printers in 1988 but stopped it just in 2007 [3]. Development of this technology raised mass production of photopolymer resins for SL. But progress wasn't still so fast. Machines were expensive and just few hundreds of them were sold.

Digital Light Processing (DLP) 3D printers work similarly to SL machines and use photosensitive polymers as build material. But instead of laser they use dynamic mask, which is generated by digital mirror device (DMD) [4]. Light beam from UV radiation generator goes through DMD chip, which can control each pixel in its matrix. It means, that we have "white" (illuminated) and "black" (non-illuminated) points on the surface. At "white" points polymerization starts, at "black" points polymer remains liquid. This way, the full layer can be solidified at the same time unlike laser stereolithography. Also using of bottom-up printing technology, which is shown on Figure 1, can

[a] Corresponding author : iaroslav.kovalenko@tul.cz

reduce volume of needed material. One of the main advantages of DLP printers is high precision and building speed. In case of laser SL printer, one point is printed in one moment of time unlike DLP printer, when a whole layer is printed simultaneously. On the other hand, duration of radiation for one layer of DLP printer is longer than for one point of laser printer. Precision depends on resolution of DMD chip and size of build area. For example, with resolution of 1900 x 1200 pixels we can reach a precision of 20 μm with size of build platform 38 x 24 mm in XY direction.

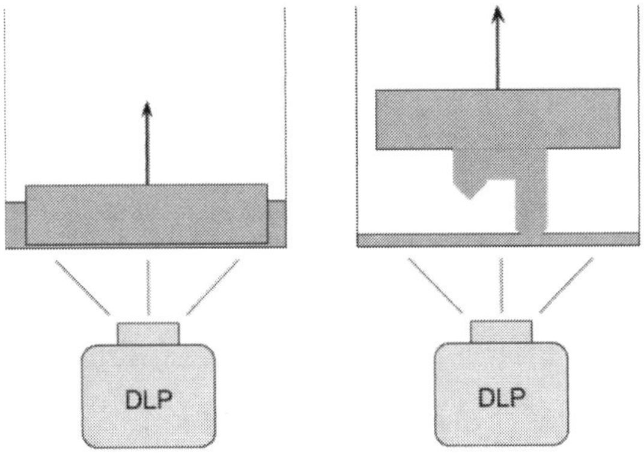

Figure 1. Schematic sketch of bottom-up 3D DLP printers

2 Photopolymer resins

Using of photopolymer solutions makes possible to add additional components, that can change properties of the solution, and consequently of built model. For obtaining of micro-structured surfaces with high precision, additional ceramic or metallic powders may be used. These additives serve for reaching of better mechanical properties of the model [5]. A lot of different methods in this area were designed and experimentally approved by TU Wien. Experiments including ceramic powders to photopolymer solution were made using alumina [6-9], silica [6], Bioglass® [7,8] and zirconia [7,9]. Different amounts of additives were added to the photopolymer solutions, mixed and used like a build material for a different types of SL printers. These solutions were used for printing of green bodies, which were sintered after.

Photopolymer solutions consist of three basic elements: monomers/oligomers, photo initiators and blockers [10]. Using of different components and different proportions of basic components makes a big variety of existing resins. Monomers/oligomers are basic components, which mostly have above 95 vol% of the solution. Photo initiators determine range of radiation, which can be used for solidifying, and curing speed. The higher percentage of initiator, the faster reaction is. Blockers define a deepness of radiation penetration and consequently have effect on layer thickness. For start of polymerization, photo initiators are exposed by light. Different initiators are sensible to specific wavelengths [11]. Typically for DLP 3D printers, resins that are sensible to UV wavelength are used. Common hand-made DLP printers for solidifying of photopolymer solution use office projectors. They are cost-effective, but mostly they have UV wavelength as unwanted output. In our research we compare radiant energy, radiated the experimental surface, from common office DLP projector and industrial projector with UV LEDs.

3 Experimental setup

For this test we used widespread office projector Acer H6510BD and well-known industrial projector DLP LightCrafter E4500 (Texas Instruments Inc.) with replaced lamp for UV LED (EKB Technologies Ltd.). Main properties of compared projectors are presented in Table 1.

Table 1. Main properties of tested projectors

Name	**Acer H6510BD**	**LightCrafter E4500**
Native Resolution	1920 x 1080 (FHD)	1280 x 800 (WXGA)
Aspect Ratio	16:9	16:10
Size (H x W x D)	80 mm x 220 mm x 260 mm	48 mm x 140 mm x 152 mm
Lamp Wattage	210 W	7.5 W
Lamp Wavelength	Visible spectrum	385 – 405 nm
Throw Ratio	1.15:1 - 1.50:1	1.2:1
Offset	29 %	0 %
Min. Image Size	53.3 mm x 30 mm	20.4 mm x 12.7 mm
Min. Working Distance	50 mm	25 mm

GaAsP Photodiode was used as a sensor for measurement of radiant fluence. It has a spectral response in the range of 280-400 nm with the peak near 370 nm. This type of sensor was chosen because it is sensible to wavelengths of light that are used for solidifying of common photopolymers.

Projector was fixed at the mobile platform, therefore distance between lens and projected area was able to change. Two different distances were chosen for measurement. These distances were set up to reach two different illuminated areas. Area of illuminated part of the platform has influence on precision of a printed model.

A sensor was installed on the console in front of projector on the projected area. Duration of all measurements was set to 200 seconds. Measurement started in reference point. After finishing of measurement in the first point sensor with a frame was moved to the second point. This routine was repeated for every chosen point on illuminated area.

Figure 2. Image of measurement process of LightCrafter

Measurement for LightCrafter projector, which is shown on Figure 2, was provided for two areas with different distance between lens and projected area. It was made for getting information about influence of distance on radiant fluence. For Acer projector the one experiment with constant distance was enough, because during test radiant energy had relatively low level and providing of second experiment was useless, as soon as no additional information could be provided.

4 Results and discussion

Data from a measurement are presented below. Figure 3 shows distribution of radiant energy for LightCrafter projector on printed area 80 x 50 mm. In this case one pixel (one mirror of DMD chip) has a size 62.5 x 62.5 μm on the projected surface. Precision of the printing on this area will be equal to a size of one pixel. Bottom side of Figure 3 shows 2D wire graph of radiant fluence. Maximal measured value for this case is equal to 156 mJ/cm^2 that corresponds to 100%. Bottom part of figure shows, that the peak of radiant fluence is not in the center of printed area. Also it is important to note that the value of radiant fluence is significantly smaller in the corners of printed area. It signifies that process of solidification will not be even on the whole area. This effect is displayed better during printing with short time intervals of radiation. In this case some areas got not enough radiant energy and did not solidify properly. It can cause a damage of whole model, as it is shown on Figure 4. That's why effective printing space is smaller than illuminated area that is shown on Figure 3.

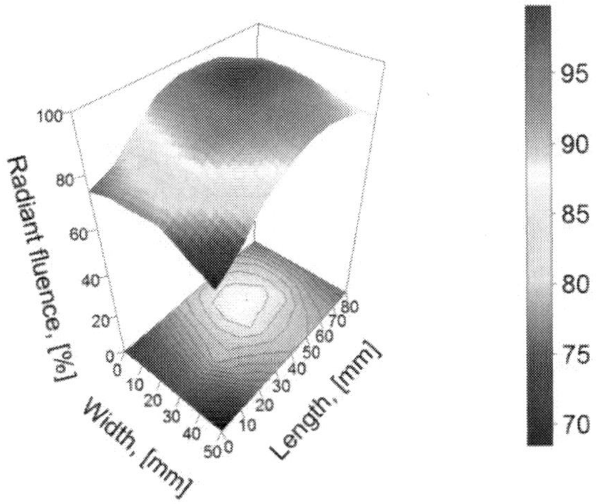

Figure 3. Relative radiant fluence of LightCrafter projector with precision of 62.5 μm

Figure 4. The damaged printed model

Increasing of effective printing space up to illuminated area is possible by rising a time of radiation for each layer. For this projector overheating of the platform was not noticed, because most of consumed energy were transformed to light beam in UV range.

Difficulties with overheating are common for not UV projectors. Visible spectrum, which is generated by Acer projector, has a low amount of UV light. It means, that big amount of energy, which is consumed by the projector, is used not for solidification of build material, but for heating of the platform, model and photopolymer solution. It can cause internal stress in built model, which may lead to mechanical defects, such as bending during or after printing. Also some parts of the model may remain non-solidified, then the resulting model becomes damaged, similarly to the model, shown on Figure 4.

Sometimes built models can unstick from the platform in the corners or even all over the entire surface. It can cause damaging of a tank with photopolymer. In this case using of manually added support structures can be effective. But in the case of DLP printers, it is not always useful and take more time for preprocessing.

In Figure 5 results of measurement of radiant fluence for the LightCrafter projector but on the different printing area are shown. In this case printing area is 40 x 25 mm with precision of 31.25 μm. Distance between projector and sensor is lower, that's why amount of radiant energy is considerably higher, than in the first case. Consequently, unevenness is higher as well (in comparison to measurements on the larger area). Maximal radiant fluence in this case is 760 mJ/cm^2. Measurement was provided in the same number of points.

 DOI: 10.1051/matecconf/20178801001

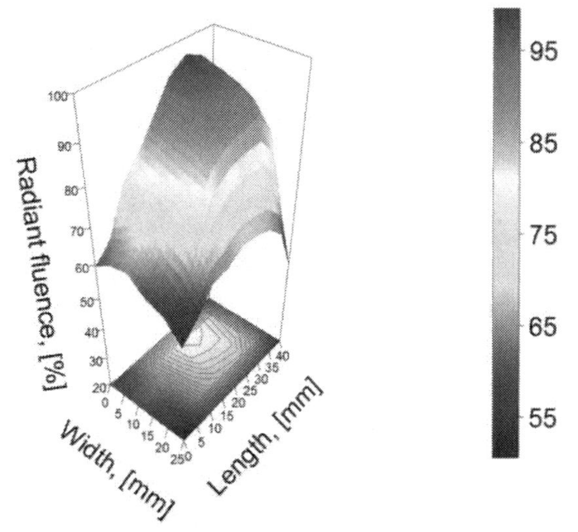

Figure 5. Relative radiant fluence of LightCrafter projector with precision of 31.25 μm

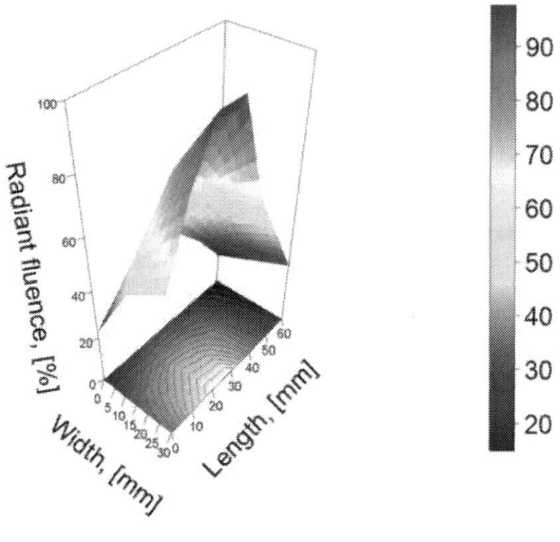

Figure 6. Relative radiant fluence of Acer projector with precision of 31.25 μm

Figure 6 shows relative value of radiant fluence for the Acer projector. Printing area for this measurement is larger, than for the second case of LightCrafter projector (Figure 5). It is 59.4 mm x 33.8 mm. Precision remains the same as in previous case, 31.25 μm. It is caused by bigger DMD chip in Acer projector with resolution 1920 x 1080 mirrors. Maximum value of radiant fluence is 31 mJ/cm^2. The bottom part of Figure 6 shows, that the peek radiant energy is directly at the verge of printing area. It is caused by the offset of the Acer projector. It causes nonuniform solidifying effect at some distance from center and decreases effective printing area in comparison with a LightCrafter projector.

5 Conclusions

Common widespread Acer DLP projector generates just 4% of radiant energy in comparison with industrial LightCrafter projector. Measurements showed that the peak of radiant energy for the LightCrafter projector is closer to the center of printing area as compared to the Acer projector. This option can rise effective printing area, even in comparison with Acer projector which has higher resolution of DMD chip and larger printing area. And the last, but not the least important parameter of tested projectors is energy consumption, which is for Acer projector higher, as compared to the LightCrafter projector.

All these advantages of industrial projector not only make 3D printer more effective, but also improve quality of printed models. Reliability of 3D printer can be increased, since an undesired heating of the platform can be weakened or even illuminated.

Acknowledgement

The research reported in this paper was supported by targeted support for specific university research within the student grant competition TUL (Project 21071 – Development and prototype production of compact DLP 3D printer).

References

1. J.M. Waller, B.H. Parker, K.L. Hodges, E.R. Burke, J.L. Walker, E.R. Generazio, *Nondestructive Evaluation of Additive Manufacturing* (Langley Research Center, Hampton, 2014)
2. T. WohlDer, T. Gornet, *Wohlers Report 2016 : 3D Printing and Additive Manufacturing State of the Industry*, (Wohlers Associates Inc., USA, 2016)
3. C.W. Hull, *Apparatus for production of three-dimensional objects by stereolithography*, U.S Patent 4575330 (1986)
4. L.J. Hornbeck, *Frame addressed spatial light modulator*, U.S. Patent 4615595 (1986)
5. A. de Blas Romero, M. Pfaffinger, J. Mitteramskogler et al. Int. J. Adv. Manuf. Technol. (2016) doi:10.1007/s00170-016-8856-1
6. M.L. Griffith, J.W. Halloran, J. Am. Ceram. Soc., **79**, 10 (1996)
7. R. Gmeiner, G. Mitteramskogler, J. Stampfl, Int. J. Appl. Ceram. Technol., **12**, 1 (2015) DOI:10.1111/ijac.12325
8. R. Felzmann, S. Gruber, G. Mitteramskogler, P. Tesavibul, A.R. Boccaccini, R. Liska, J. Stampfl, Adw. Eng. Mat. **14**, 12 (2012) DOI: 10.1002/adem.201200010
9. G. Mitteramskogler, R. Gmeiner, R. Felzmann, S. Gruber, C. Hofstetter, J. Stampfl, J. Ebert, W. Wachter, J. Laubersheimer, Add. Manuf., 1-4, (2014)
10. R. Holtrup, *Design and construction of a multi-material 3D DLP printer*, (University of Twente, Netherlands, 2015)
11. V. Bernabé-Zafón, M. Beneito-Cambra, E.F. Simó-Alfonso, J.M. Herrero-Martínez, Journal of Chromatography A, **1217** (2010)

Enhancement the Armor Shielding Properties of CF/epoxy Composite by Addition Nanoparticles of Magnetic Iron Oxide

Hany Fouda[1,a], Khalid Ghith[2] and Lin Guo[1]

[1]*Beihang University of Aeronautics and Astronautics, Beijing, china.*
[2]*Military technical college, Cairo, Egypt.*

Abstract. In the present investigation, we prepared two types of CF composites. The first prepared composite sample was CF/epoxy resin composite and the second was CF/epoxy resin / with a different weight ratio of magnetic iron oxide. Magnetic iron oxide was prepared by co-precipitation method, with particle sizes measured in range 25:35 nm. The resistance to penetration of high kinetic energy projectile of the prepared composite sample was measured and It was found that addition of 5% nano-particles of magnetic iron oxide to composite material sample decrease the residual velocity of projectile penetrating it by 9%.i.e increasing resistance of the sample to penetration of high kinetic energy projectile.it was found that the Resistance to penetration of sheet of composite material sampleC4 of weight=40.32kg to projectile 7.62×39 mm AP at distance 15m equivalent to resistance of steel sheet of weight =54.6 kg at distance 200m.Resistance to penetration of sheet of composite material sampleC4 to projectile 7.62×39 mm AP at distance 10m equivalent to the resistance of high-quality steel sheet(steel4340) of weight=47.85 kg at distance 25m.

1 Introduction

Carbon fiber CF considered as one of the most important fillers for polymeric matrix composite materials due to its unbelievable properties, such as high thermal stability, high specific strength and stiffness, performance to weight ratio, high conductivity, corrosion resistance and self-lubrication. Most significantly, use of CF allows a reduction in weight of the equipment's or vehicles due to its high strength to weight ratio. CF reinforced polymer composites used for aerospace efficacies, wind turbines applications in automotive energy systems, fuel cells, offshore– deep sea drilling platforms, compressed gas storage, turbo machinery and transportation, antistatic and electromagnetic shielding materials[1-3]. Liu and Kumar have studied the existing progress of carbon fiber structure, fabrication and properties including the introducing of nano-tubes in precursor fiber to improve the mechanical properties [4-6]. However, the essential mechanical characteristics of these composites such as toughness, longitudinal and transverse strength limited by the intrinsically poor interfacial adhesion between polymer materials and reinforcing CF surface. It is a long existing critical issue needed to be solving for ensuring the continued development of CF reinforced polymer composites for potential advanced composites applications [4-9].

[a] Corresponding author: hanyfouda@buaa.edu.cn

Motivated by this, many researchers focused on probing an understanding the physicochemical interaction at the fiber/matrix interface. For achieving strong interfacial bonding, the adequate level of Vander Waals and hydrogen bond forces between the matrix and CF are required during composite processing.

In addition, the fiber/matrix interfacial adhesion energy should be higher than the cohesion energy of the matrix. The modifications applied to CF structure made a big difference in improving the mechanical properties of high performance polymeric composites but the focus to control the fiber/matrix interfacial properties is still a major task [8-9].

CF, from a structure has crystallized graphitic basal planes with non-polar surface. The chemical inertness due to the existence of high temperature treatment step [carbonization/graphitization] during manufacturing, surface lipophobicity, and excessive smoothness and fewer adsorption properties of CF lead to weaker bonding with the matrix materials.

As remedy for CF inertness, modifications at fiber surface needed to execute strong fiber/matrix interfacial adhesion for effective stress transfer at the interface [10].

Epoxy resins are excellent and high-performance thermosetting resins, which display a unique combination of properties. Epoxy resins are one of the most versatile polymers with uses across an enormously wide variety of industries. They are composed of polymeric molecules that are converted to a solid by a chemical reaction. The ability to be transformed from a low-viscosity liquid (or thermoplastic state) into a tough, hard thermoset is the most valuable single property of epoxy resins. An epoxy system physically comprises two essential components: a resin and a curative. The curative causes the chemical reaction, which turns the epoxy resin into a solid, cross linked network of molecules. This polymer is called a thermoset polymer structure with high cohesive strength and adhesion properties because, when cured, it is relatively unaffected by heat and irreversibly rigid. However, the term epoxy can also be used to indicate an epoxy resin thermoplastic or cured state [11-14].

Many studies about the increasing the mechanical properties of CF/epoxy resin composite were applied, Mainly by the addition of nanoparticles materials as second filler to increase the bond between CF and epoxy resin such as carbon nano fiber or carbon nanotube. Carbon nanotubes (CNTs) have been widely used in enhancing composites for its excellent mechanical properties, large specific surface area, as well as good compatibility with polymer [8]. The method of introducing CNTs on the surface of fibers to enhancement the interfacial properties of composites has been a hot topic.

Jianwei Zhang, Su Jua, Dazhi Jiang, Hua-Xin Peng, enhanced Mechanical properties of carbon fiber (CF) reinforced epoxy composites with addition a small amount of multi-walled carbon nanotubes (MWCNTs). 1 wt % MWCNTs were well dispersed in the epoxy matrix and the CF prepregs were manufactured using the technique of fiber filament wind. The hybrid composites of MWCNT/CF/epoxy were fabricated by laying-up of fiber prepregs. Interlaminar shear strength (ILSS), tensile properties and microstructures of the composites were investigated. The results show that the addition of MWCNTs into the composites increased and enhancement the tensile properties and ILSS with much reduced disparity. Enhancing mechanisms behind these were elucidated [15].

In this research, we use the same concept to add nanoparticle filler but with a different point of view. In this investigation, the effect of the addition of nano particles of magnetic iron oxide into the resistance of CF/epoxy resin composite to penetration of high kinetic energy projectile was studied.

2 Experimental work

2.1 Materials

The carbon fiber used is PAN carbon fiber made from a polyacrylonitrile precursor has moderate strength; moderate modulus and carbon fiber yarn contain 3000 filaments obtained from Russia. The epoxy resin used in this study is a two component epoxy system consisting of (EPON 828) cured with polyamide (versamide 125) hardener. This two component system obtained from HEXION Company, France .epoxy processed in proportions of 1:1 by weight. Carbon fiber impregnated with phenol formaldehyde resin obtained from Russia. The nano-scale fine particles used are magnetic iron oxide [Fe3O4] prepared by precipitation method.

2.2 Nano-particles used of magnetic iron oxide [Fe_3O_4]

Super –paramagnetic magnetite (Fe_3O_4) were produced by a co-precipitation method from the aqueous solutions of $FeCl_2$ and $FeCl_3$ using a NaOH or NH_4OH solutions [17-18].By applying the following standard procedures. First, prepare 0.11M $FeCl_2$ and 0.2 M $FeCl_3$ in 1 liter distilled water and put the solution in a reactor and stirred vigorously. Then put NH_4OH solution in separating funnel then add ammonia solution drop by drop to iron chloride solutions with strong stirring till black precipitate formed then added an excess amount of NH_4OH solution and then agitate for 1 hour. After 1 hour we put the reactor in the water bath at 80^0C with strong stirring for 120 minutes. Then we stop stirring and lit the suspension particles to precipitate. After that we add distilled water to the reactor to wash particles and leave it till precipitate then we poured some water without the particles. We repeat the previous procedure several times till we washing iron oxide particles and still the iron oxide particles suspending in few drops of remaining water. Then we put the reactor in the furnace at 60^0C for 2 days to evaporate the liquid and achieved on the iron oxide particles. Finally, we weight the particles and put it in the suitable bottle.

2.3 Preparation of epoxy samples

Epoxy resin system consists of two parts A (EPON 828) and B hardener (versamide 125), part A was carefully weighed and a stoichiometric amount of part B was added, then stirring using magnetic stirrer for 30 min with low velocity to prevent air trapping. When addition nano-scale fine particles to composite material we added it first to part A of polymer and mixed together in the suitable beaker for 10 min for homogeneity [13].

2.4 Composite fabrication

Composite material based on CF/epoxy resin and was prepared by hand lay-up technique. Brush and roller were used to help the impregnation of fiber. The epoxy resin was prepared first as in the previous paragraph 2.2. The epoxy was brushed on the surface of CF. the epoxy-brushed fiber tape was carefully stacked up and aligned together layer above layer then compressed adding load 50 Kg over the upper surface and left at room temp for 24 hrs [11].

2.5 Determination of particle size of prepared magnetic iron oxide

The particle size of fumed silica and prepared magnetic iron oxide can be measured by several techniques as particle size analyzer (zeta sizer) or can be measured by High Resolution Transmission Electron Microscope. In our research, we used the last method for determination of particle size which was high resolution transmission electron microscope (TEM), type JEOL JEM 2010 FAS TEM, purchased from JAPAN. The used TEM specified by ultra-high resolution point-point reached to 194 nm and lattice resolution reached to 14nm.

2.6 Penetration tests

DOI: 10.1051/matecconf/20178801002

Composite materials generally and CF composite especially are suitable for use in applications involving high velocity impact because they combine lightness with high strength. However, the impact of composite materials is a very complex process. As a consequence of this CF, matrix and interface properties are still optimized by trial and error using ballistic experiments. So the main objective of the experimental work was to study the penetration of 9mm and 7.62 * 39 mm projectiles into a set of laminated CF /epoxy composites targets and body armors using the experimental facilities of the shooting ranges, Weapons and Ammunition Researches Sector, Technical Research Center (T.R.C), Egypt. Ballistic tests were performed to determine the projectile impact and post-perforation velocities for the prepared targets from CF/epoxy composites with thickness 7mm and 35 mm. Projectile arrivals were detected using the muzzle velocity radar measuring system, Model 2605R. The Departure of the projectile was detected using the velocity measuring device, Model HPI B571 optical target system.

In this test, we tested the impact resistance of the prepared composite samples against small projectile as 9mm projectile and 7.62× 39 mm armor piercing (AP) projectile in order to explain the application of prepared composite as light armor. The composite samples were tested against 9mm projectile from 10m distance, while the another sample was tested against 7.62×39mm AP projectile from 10m distance [16].

2.6.1 Penetration setup

The penetration experiments were performed in the ballistic shooting range, which had provisions for the measurement of projectile impact and post-perforation velocities respectively, for prepared CF/epoxy composites targets. As shown in figure 1, the penetration setup mainly consists of: ballistic rifle, impact velocity and post perforation velocity frames with their respective electronic measuring instrument, and target mount.

By using the simulation program (AUTODYNE program) and calculations of absorbed energy and kinetic energy we compare the prepared sample with resistant of steel against the same projectile.

2.6.2 Setup of penetration test

To apply penetration test we should have penetration set up consist of the following parts:
A. Ballistic rifle used for penetration test

They are standard barrels, caliber 9 mm and caliber 7.62mm used for the particular purpose of the ballistic measurements of 9 × 19 and 7.62 × 39 bullets. The barrel was mounted horizontally during the whole tests. The 9 mm and 7.62 mm barrels were fixed on a rigid stand as shown in Figure 1, which in turn, was mounted on a concrete base by supporting studs. The barrel direction can change in elevation and traverse.

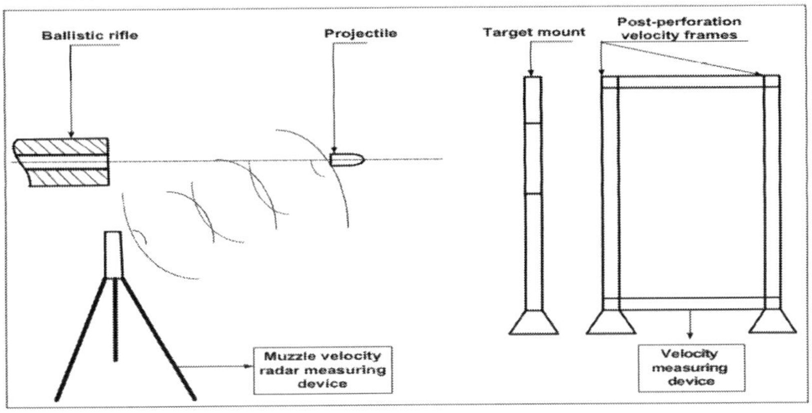

Figure 1. Scheme of ballistic set-up.

The test was carried out according to the following steps:
Penetration tests were performed in the shooting range according to the following steps:

1. The muzzle velocity measurement system, Model 2605R and the residual velocity measuring device, Model HPI B571 optical target system were first switched on. The computers were switched on and the software was activated.
2. All electronic instruments used were reset again to be ready for measurements
3. The prepared target was fixed in place.
4. The projectile giving a predetermined impact velocity was chosen and loaded into the barrel chamber. This was only allowed after evacuating the shooting range from personnel.
5. The safety door was closed and firing performed electrically from the control room. The displayed impact and residual velocities on their respective PC screens were recorded.
6. The breech was unlocked and empty cartridge case extracted.
7. The shooting point on the target was marked.
8. For firing another shot, the steps from 2-7 were repeat

2.7 Samples codes and compositions

The codes and composition of the prepared composite samples used in this research illustrated in table 1.

Table 1. Details of composition of each formulation.

Formulation	Composition
C1	[CF+ epoxy resin]composite with thickness 7 mm
C2	[CF+ epoxy resin+ 2.5% iron oxide] composite with thickness 7 mm
C3	[CF+ epoxy resin+ 5% iron oxide] composite with thickness 7 mm
C4	[CF+ epoxy resin+ 5% % iron oxide] composite with thickness 35 mm

3 Result and discussion

3.1 Particle size determinations of magnetic iron oxide particles

The particle size of prepared magnetic iron oxide can be determined from the figure 2, which show the TEM picture for the prepared magnetic iron oxide. The particle size of prepared magnetic iron oxide was in the range of 25:35 nm.

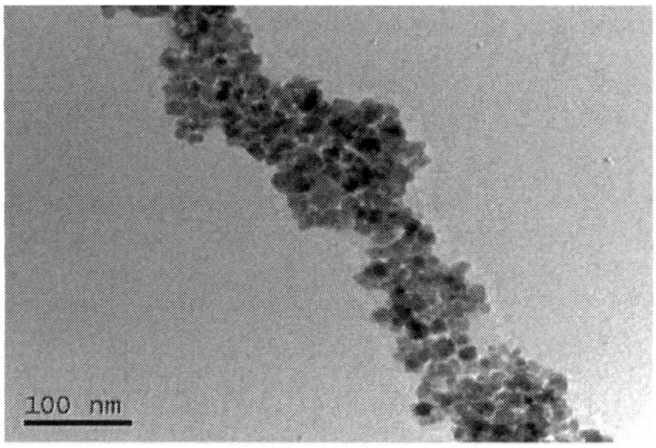

Figure 2. Particle size of magnetic iron oxide.

3.2 Resistance of prepared composite material to penetration of high kinetic energy projectile

In this part, the resistance of prepared composite material samples to penetration against high kinetic energy projectile in order to use in armor applications will be discussed. The resistance to penetration is expressed by measuring the residual velocity and energy absorbed by the sample during penetration of projectile.

3.2.1 Resistance of prepared composite samples to penetration of 9mm projectile

In this test, three composite material samples were prepared and subjected to 9mm projectile from distance 10m. The samples are C1, C2, and C3. The effect of the addition of nano-particles of magnetic iron oxide to composite material on the resistance to penetration shown in table 2.

Table 2. Resistance test result for composite material samples C1, C2andC3.

Sample code	Sample thickness [mm]	Sample density [gm/cm^3]	Impact velocity V_i [m/s]	Residual velocity V_r [m/s]	Velocity Drop ΔV [m/s]	Velocity Drop Percentag
C1	7	1.154	251.24	197.5	53.74	21
C2	7	1.12	251.7	188	63.7	25.4
C3	7	1.12	251.5	180	71.5	28.5

From Table 2, we can see that the drop in velocity due to penetration of 9mm projectile for composite sample C3 is higher than that of composite samples C2and C1 respectively. It also shows that the residual velocity of the projectile due to penetrateC3 is less than that of sample C1 by 9%.
This mean that addition of 5% nano-particles to composite material sample decrease the residual velocity of the projectile penetrating it by 9%.i.e increasing resistance of the sample to penetration of high kinetic energy projectile.

3.2.2 Resistance of prepared composite samples to penetration of 7.62×39mm AP projectile

In this test, composite material sample based on CF/epoxy resin/ 5% magnetic iron oxide was tested. The sample C4 with a thickness of 35mm was subjected to 7.62×39mm AP projectile from distance 10m.the result shown in table 3.

Table 3. Resistance test result for composite material sample C4.

Sample code	Sample thickness [mm]	Sample density [gm/cm^3]	Impact velocity V_i [m/s]	Residual velocity V_r [m/s]	Velocity Drop ΔV [m/s]	Velocity Drop Percentage %	Absorbed energy [J]
C4	35	1.12	750	290	460	61	1889.68

From table 3, it can be seen that sample C4 absorbed energy equal to 1889.68 J from the kinetic energy of the projectile at 10m which equals to 2222.5 J. From previous work which has been done in the department of weapons and ammunition M.T.C., the velocity of projectile versus distance was plotted. This plot is shown in figure 3 [16]. From this figure, the velocity at which the kinetic energy of projectile equal absorbed energy by sample C4 at distance 10m can be detected. From the figure, we found that velocity which gives the same energy was at 15m.i.e. the composite sampleR11 stopped projectile at distance 15m.

Calculations

From figure 3 at 15m velocity = 691.5 m/s

Kinetic energy of projectile at 25m $[K.E] = 0.5 \times m \times V^2$

$= 0.5 \times 7.9 \times 10^{-3} \times (691.5)^2 = 1880.595$ J

Absorbed energy by sample C4 = 1889 J

But, energy absorbed by sample C4=kinetic energy of the projectile at 15m.

So, C4 can stop the projectile 7.62×39mm AP at distance 25m.

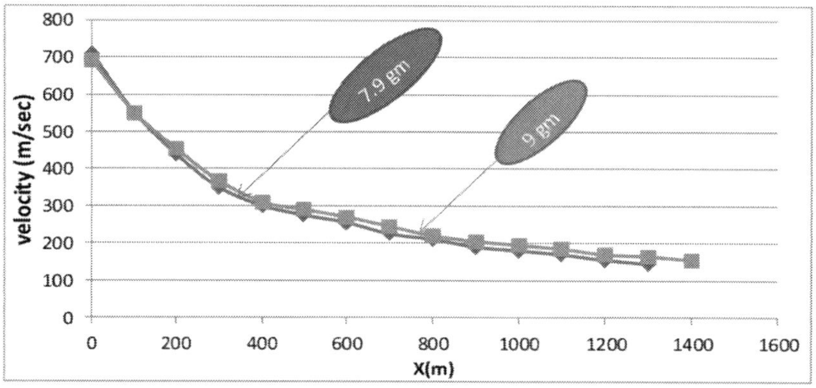

Figure 3. Velocity of 7.62×39mm AP projectile versus distance.

3.2.3 Comparison between steel and composite sample C4 to resistance of penetration

The projectile used for this comparison was 7.62×39mm armor piercing (AP). The results show that the prepared composite sample C4 can stop the projectile at distance equal 15m. However, 7mm thickness of steel can stop this projectile at a distance equal to 200m, as stated in the literature. So, the prepared composite sample gives better results than steel. In addition to the big difference of weight of the sheet of the composite of dimensions 1mx1mx35mm which is (40 kg) compared to the weight of the sheet of steel of dimension 1mx1mx7mm which is (54.6 kg) should be taken into consideration.

3.2.4 Comparison between high quality steel (4340) and composite sample C4 to resistance of penetration

The projectile used for this comparison was 7.62×39mm armor piercing (AP). The results show that the prepared composite sample C4 can stop the projectile at a distance equals to 15m. However, 6.1mm thickness of steel 4340 can stop the projectile at a distance equals to 25m, as calculated by using AUTODYNE program, and also absorbed the same amount of energy as absorbed by composite sample C4 at a distance equals to 10m. So, the resistance of the prepared composite sample to projectile penetration equals the resistance of 6.1mm of high quality steel for the same projectile. Also, the weight of a sheet of the composite of dimensions 1mx1mx35mm is 40.32 kg; however the weight of a sheet of steel of dimension 1mx1mx6.1mm is 47.6 kg.

4 Conclusion

The composite material based on CF/epoxy resin and CF/epoxy resin/nano particles of magnetic iron oxide were preparing by hand layup technique. It was found that the prepared magnetic iron oxide has particle size in range 25:35 nm. The addition of 5% nano-particles of magnetic iron oxide to CF/epoxy resin composite material improve the resistance of composite to penetration of high kinetic energy projectile expressed by a decrease of residual velocity of the projectile by 9%. Resistance to penetration of sheet of composite material sampleC4 of weight=40.32kg to projectile 7.62×39 mm AP at distance 15m equivalent to the resistance of steel sheet of weight =54.6 kg at distance 200m. Resistance to penetration of sheet of composite material sampleC4 of weight=40.32 kg to projectile 7.62×39 mm AP at distance 10m equivalent to the resistance of high quality steel sheet(steel4340) of weight=47.85 kg at distance 25m.

References

1. Schwartz MM, "*composite materials hand book*", McGraw-Hill; 1992.
2. Whitcomb JD, "*Composite materials testing and design*", ASTM; 1988.
3. S.T.Peters," *Hand book of composites*", CHAPMAN&HALL; 1998
4. Soutis C, "*Fiber reinforced composites in aircraft construction*", Prog Aerospace Sci 2005; 41:143–51.
5. Pavia JMF, Santos AN, Rezende MC, "*Mechanical and morphological characterizations of carbon fiber fabric reinforced epoxy composites used in aeronautical field*", Mater Res 2009;12(3):367-374.
6. Mallick PK, "*Fiber reinforced composites: material, manufacturing and design*", 3rd Ed. New York: CRC Press; 2008.
7. Edie DD, McHugh JJ, "*High performance carbon fibers*", In: Burchell TD, editor. Carbon materials for advanced technologies, Pergamum; 1999. p. 119–38.
8. Liu Y, Kumar S," *Recent progress in fabrication, structure, and properties of carbon fibers*", Polym Rev 2012; 52(3–4):234–58.
9. Mohit Sharma, Shanglin Gao, Edith Mäder, Himani Sharma, Leong Yew Wei, Jayashree Bijwe, "*Carbon fiber surfaces and composite interphases*", Composites Science and Technology 102 (2014) 35–50.
10. Pavia MC, Bernardo CA, Nardin M, "*Mechanical, surface and interfacial characterization of pitch and PAN-based carbon fibers*", Carbon 2000; 38(9):1323–37.
11. T. Takeichi, N. Furukawa, "*Epoxy Resins and Phenol-Formaldehyde Resins*", Reference Module in Materials Science and Materials Engineering Polymer Science: A Comprehensive Reference, pages 723-751, 2015 volume **5**.
12. Deborah D. L.Chung, "*Carbon Fiber Composites*" , Butterworth-Heinemann; 1994.
13. Ronald F.Gipson, "*Principles of composite materials mechanics*", McGraw-Hill.
14. Hui Zhang, Zhong Zang, Claudia Breidt, "*Comparison of short carbon fiber surface treatments on epoxy composites*", institute for composite materials, university of Kaiserslautern, and Erwin Schrodinger strasse 58; 2004.
15. Jianwei Zhang, Su Jua, "Dazhi Jiang, Hua-Xin Peng, "*Reducing dispersity of mechanical properties of carbon fiber/epoxy composites by introducing multi-walled carbon nanotubes*", Composites: Part B 54 (2013) 371–376.
16. Walled Magdy," *production technology of composite materials based on polymeric matrices*" Master, MTC, Egypt, 2004.
17. Xueai Li, Xijiang Han, Yanjiang Tan, and Ping XU, "*Preparation and microwave absorption properties of Ni-B alloy- coated Fe3O4 particles*", Journal of alloys and compounds, pp. 240-245, 2007.
18. Xiang Li , Haibin Yang, Wuyou Fu, Shikai Liu, Hongyang Zhu, and Xiaofen Pang," *Preparation of low-density super paramagnetic microspheres by coating glass micro balloons with magnetic nanoparticles*", Elsevier, Materials science and Engineering, pp. 38-43, Vol.2B 135, 2006.

A Study on Giant Magnetoimpedance Effect in VITROVAC 6025Z Micro-patterned Ribbons

Zhen Yang[1,a], Tianqi Fan[2] and Chong Lei[2]

[1] *School of Physics and Electronic Engineering, Xinyang Normal University, Xinyang 464000, People's Republic of China*
[2] *Key Laboratory for Thin Film and Microfabrication of the Ministry of Education, Department of Micro/Nano Electronics, School of electronic information and electrical engineering, Shanghai Jiao Tong University, Dong Chuan Road 800, Shanghai 200240, People's Republic of China*

Abstract. In this paper, micro-patterned cobalt-based amorphous ribbons with different structures are fabricated by MEMS technology, including bonding, lithography and electroplating. The commercial amorphous ribbons (VITROVAC 6025Z) are purchased from VACUUMSHMELZE® company. The structures of micro-patterned ribbons are designed for meander shape for different turns (one turn, two turns and three turns). The effect of external magnetic field orientation and turns on GMI ratios were studied in the frequency range of 1-40 MHz. It is found that GMI ratios almost all present negative value, which is related to the material property, namely a large coercive force. Compared with the GMI ratios obtained at transverse external magnetic field, the maximum negative GMI ratio obtained at longitudinal external magnetic field is higher. The value of -69.93% is obtained at a current frequency of 20 MHz and the magnetic field of 150 Oe. With the increase of the number of turns, the maximum positive GMI ratio increases from 0.5 % for the one turn sample to the 2.8% for the sample with three turns. The maximum nagative GMI ratio is obtained by three-turns sample. The GMI character of VITROVAC 6025Z micro-patterned ribbon is significant and the microsensor is expected to be used in biological sensing field.

1 Introduction

Giant magnetoimpedance (GMI) effect is defined as a large change in the ac complex impedance of a soft ferromagnetic conductor upon the application of a dc magnetic field. It has drawn an intensively research activity by reason of promising technological applications based on the GMI phenomenon since its discovery [1, 2]. Among various kinds of existing GMI materials, soft ferromagnetic amorphous ribbons are considered attractive candidates for making all kinds of GMI-based magnetic sensors [3-5], because they possess a stripe-like domain structure along the transverse direction, which is favourable for achieving GMI effect with a high degree of field sensitivity.

In recent years, the GMI effect of micro-patterned ribbon with tortuous structure is extensively investigated. Because meander-shaped micro-patterned ribbon possesses high GMI ratio compared that of the ribbon with single strip structure [6-7]. Cobalt-based amorphous alloys, e.g., Vitrovac,

[a] Corresponding author : zhc025@sjtu.edu.cn

exhibit a high GMI level, a high sensitivity to an applied magnetic field [8], which makes them promising GMI materials.

So, in this context, we employ micro electro-mechanical system (MEMS) technology to fabricate micro-patterned cobalt-based amorphous ribbons (VITROVAC 6025Z) with tortuous shape for different turns. The GMI ratio of the ribbons are investigated at different magnetic fields (1-150 Oe) and frequencies (1-40 MHz). The influence of external magnetic field orientation and turns on GMI ratios are observed. The results show that the GMI ratios obtained at longitudinal external magnetic field is higher than that obtained at transverse external magnetic field. And with the increase of the number of turns, the maximum of the GMI ratio increases from 31.2 % for the one turn sample to the 69.93% for the sample with three turns.

2 Experimental procedures

2.1 Soft magnetic materials

In this section, the micro-patterned GMI sensor elements are made of a cobalt-based commercials amorphous ribbon (VITROVAC 6025Z). The thickness of ribbon is 25 μm. The main physical and magnetic properties of the ribbons are listed in Table 1.

Table 1. The main physical and magnetic properties of commercials ribbon.

Parameters	**6025Z**
Saturation induction B_S (T)	0.55
Saturation magnetostriction constant	$-3x10^{-7}$
Maximum permeability μ_{max}	$1.03x10^{5}$
Coercivity H_c (A/m)	2.84

2.2 Fabrication of micro-patterned ribbons

The micro-patterned ribbons with tortuous shape (one-turn, two-turn, and three structures) (as shown in Figure. 1) were designed with width of 500 μm and length of 10 mm. The intervals of tortuous shape were 60 μm. Each 'n' shape in the meander structures was called one turn. Before MEMS processing, the ribbons were mechanical polished. After mechanical polishing, the thickness of the ribbon was measured by surface profiler, and was 15 μm.

The ribbons with different structures were fabricated by bonding, photolithography, chemical etching and electroplating methods. The manufacturing process consisted of the following steps:

I) Cleaning substrate: At first, the glass substrate were cleaned by ultrasonic in acetone solution and alcohol in turn for 10 min, and then the substrates were rinsed with deionized water and alcohol, at last it was dried using a stream of nitrogen gas;

II) Bonding ribbon: the ribbon was bound on clean glass substrates using an epoxy adhesive, during binding we should try to ensure tight bonding and no air between the ribbon and the substrate;

III) Lithography: photoresist was spun on ribbon with the thickness of 15 μm, and then the ribbon was patterned with a mask by UV lithography;

IV) Chemical etching: the ribbon was etched in acidic mixed solution (HNO_3: HCl: H_2O_2: H_2O= 1:2:4:8) for 5 min. The error of ±2.5 μ m was found after etching process and it was considered to be acceptable. At last, the photoresist was removed;

V) Electroplating Cu electrode: Before electroplating, firstly, polyimide was filled the space between neighbor segments in order to avoid the adverse effects of the nonuniform surface, and then

Cr/Cu seed layer was deposited on patterned ribbon by RF sputtering, at last Cu electrodes with thickness of 20 μm were electroplated;

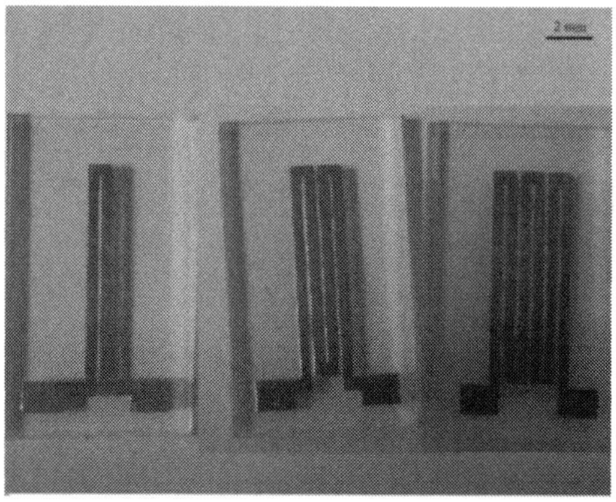

Figure 1. The top view of fabricated micro-patterned Co-based amorphous ribbon.

2.3 Measuring methods

The GMI effects in micro-patterned ribbon were measured by an impedance analyzer (HP4194A) in the frequency range of 1-40 MHz with the constant current amplitude of 10mA. An external magnetic field (H_{ex}) of 1-150 Oe was applied along transverse and the longitudinal direction of the sample. Magnetic field is got from a permanent, the magnitude of that was controlled through changing the distance between the permanent and micro-patterned ribbon. The permanent slowly closed to the ribbon along the central axis of ribbon. Magnetic fields at different locations were measured strictly by a magnetometer, so the different marked locations standed for different magnetic fields [6].

Figure. 2 shows schematic diagram of (a) transverse and (b) longitudinal magnetic field. The GMI ratio is defined as: GMI ratio (100 %) =100 %× $[Z(H)-Z(H_0)]/Z(H_0)$, where $Z(H)$ and $Z(H_0)$ are the magnetoimpedance with and without magnetic field respectively.

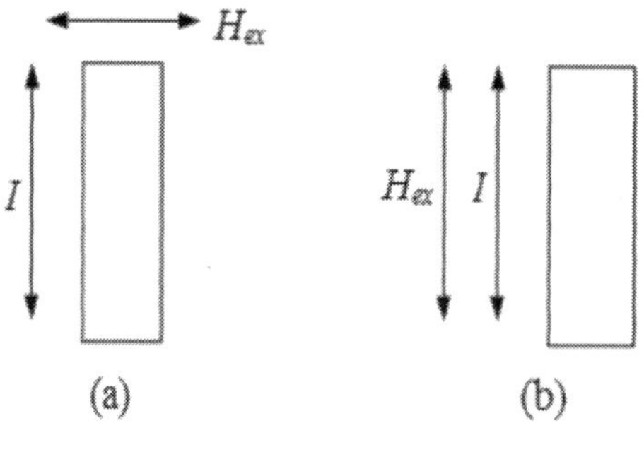

Figure 2. Schematic diagram of (a) transverse and (b) longitudinal magnetic field.

3 Results and discussion

3.1 Effect of external magnetic field on GMI

Figure. 3 shows field dependence of the GMI ratio of the tortuous-shaped ribbons at different frequency with the transverse fields. In sample testing, an external magnetic field of 1-150 Oe is applied along the transverse direction of the meander ribbon and is parallel to the transverse anisotropy direction of the sample.

The positive GMI ratio is very small and appear at small external magnetic field (1-5 Oe). When the external magnetic field is greater than 5 Oe, the GMI ratio all presented negative value. The maximum negative GMI ratio is obtained at frequency f = 20 MHz. On the whole, with increasing the applied magnetic field H_{ex}, the magnitude of the GMI ratio first reaches a slight sharp peak and then gradually decreases to zero and even negative values. The negative values become more and more big with increasing H_{ex}. In the measurement range, the maximum negative GMI ratio is -41.6% obtained at H_{ex} = 150 Oe and frequency f = 20 MHz. Our preliminary forecast a maximum negative GMI peak will appear quickly.

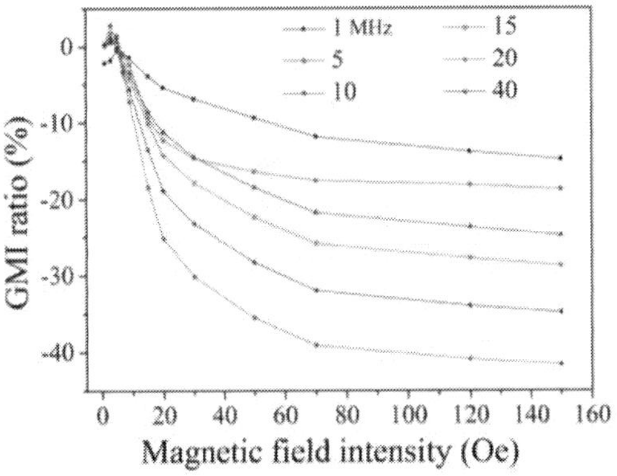

Figure 3. Field dependence of the GMI ratio of the tortuous-shaped ribbons at different frequency with the transverse fields.

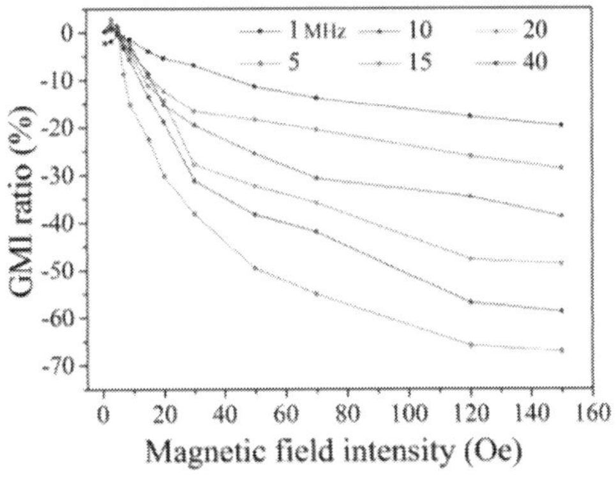

Figure 4. Field dependence of the GMI ratio of the tortuous-shaped ribbons at different frequency with the longitudinal fields.

Figure. 4 shows field dependence of the GMI ratio of the tortuous-shaped ribbons at different frequency with the longitudinal fields. Compared with figure.3, in the measurement range, the maximum negative GMI ratio is -69.93% obtained at H_{ex} = 150 Oe and frequency f = 20 MHz. The phenomenon of GMI change has been observed by Sommer [9]. Sommer indicates that a longitudinal GMI spectrum of soft magnetic materials with high transverse permeability has the typical characteristic. The GMI curves in Figure. 4 agree well with this typical characteristic, which can be explained by the magnetization rotation model and it goes as follows: the rotational magnetic permeability related to the GMI ratio increases with the increase in Hex and reaches a maximum value as H_{ex}= H_K, where H_{ex} is perpendicular to the transverse anisotropy direction of the sample and H_K is the anisotropy field. Thereafter, the permeability decreases with H_{ex} and the GMI ratio gradually drops. The difference between the longitudinal and the transverse GMI effect is due to the two main reasons: one is the easy axis deviates from the transverse direction of the sample; the other is the demagnetizing factor $D_\perp$ effect on the longitudinal permeability. Compared with the previous research [6], we design more appropriate size. The error of ±2 μm was found and smaller than previous error (±5.5 μm) after etching process, and it was considered to be acceptable. The small error leads to high GMI performance.

3.2 Effect of turn on GMI

Figure. 5 shows field and frequency dependence of the longitudinal GMI ratio of the tortuous-shaped ribbons with different turns. And the maximum positive GMI ratios obtained at a field of 3 Oe increases from 0.5 % for the one turn sample to 2.8 % for the sample with three turns. The maximum negative GMI ratios were obtained at different frequencies for samples with different turns. The maximum negative GMI ratios of one-turn, two-turns and three-turns samples were -34.8%, -50.9% and -66.9%, obtained at 4.5 MHz, 10.5 MHz and 20 MHz respectively.

The results show that the ribbons with tortuous shape of three turns acquire the maximum positive GMI and the maximum negative GMI ratios at a relatively small magnetic field and a higher frequency. This is related to the material property, namely a large coercive force. Besides, three-turns samples has more complex mutual inductance. When the ac current flows through the tortuous ribbon, just as for the flux in each longitudinal and transverse straight line in a plane inductance, the fluxes will correlate with each other by mutual inductance of the magnetic chains. The more turns of the meander structure, the more intense the interaction becomes. As a result, these effects lead to the increase in L_m and Z. More turns mean to stronger mutual inductance and higher GMI ratio, but big size is not suitable for miniaturization of microsensor.

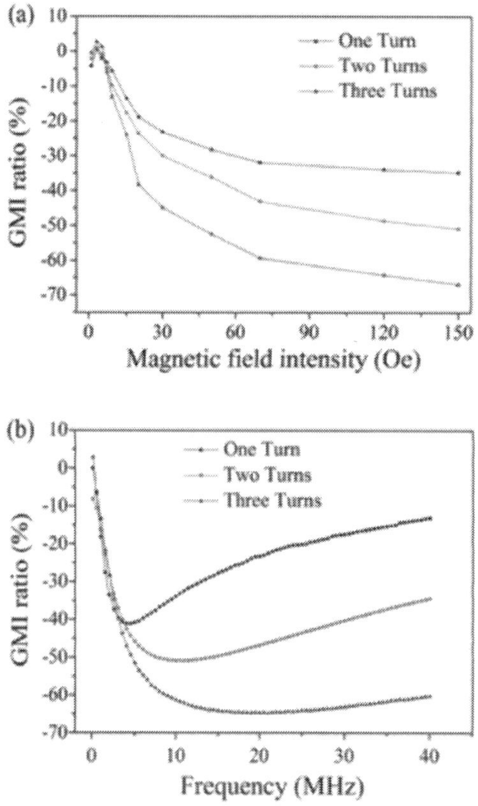

Figure 5. (a) Field and (b) frequency dependence of the GMI ratio of the tortuous-shaped ribbons with different turns.

4 Conclusion

In conclusion, the GMI ratios of VITROVAC 6025Z micro-patterned ribbons with meander structure for different turns almost all present negative value, which is related to the material property, namely a large coercive force. The small positive GMI ratio only appear when the external magnetic field is in the range of 1-5 Oe. The effect of external magnetic field orientation and turns on GMI ratios were studied in the frequency range of 1-40 MHz. It is found that the maximum negative GMI ratios obtained at longitudinal external magnetic field is higher than that obtained at transverse external magnetic field. The value of -69.93% was obtained at a current frequency of 20 MHz and the magnetic field of 150 Oe. With the increase of the number of turns, the maximum positive GMI ratio increases from 0.5 % for the one turn sample to the 2.8% for the sample with three turns. The maximum nagative GMI ratio was obtained by three-turns sample. The GMI character of VITROVAC 6025Z micro-patterned ribbons is significant and the microsensor is expected to be used in biological sensing field.

Acknowledgments

This work was supported by The National Natural Science Foundation of China (No. 61273065),

References

1. L. Kraus, M. Vázquez, M.Knobel, K. H. J. Buschow, Nort-Holland (2003).
2. V. Zhukova, A. Chizhik, A. Zhukov, A. Torccunov, V. Larin, González, IEEE Trans. Magn. 38, 3090 (2002)
3. A. Chaturvedi, K. Stojak, N. Laurita, P. Mukherjee, H. Srikanth, and M. -H. Phan, J. Appl. Phys. 111, 07 (2012).
4. Y. Zhang, J. Dong, E. X. Feng, C. Q. Luo, Q. F. Liu, J. B. Wang, Chin. Phys. Lett. 30, 037501 (2013).
5. L. González, J. Bonastre, T. Sánchez, J. D. Santos, M. L. Sánchez, A. Chizhik, L. Domínguez, M. Ipatov, V. Zhukova, A. Zhukov, J. González, J. J. Suñol, B. Hernando, IEEE Trans. Magn. 48, 4375 (2012).
6. Z. Yang, J. Lei, C. Lei, Y. Zhou, T. Wang, Appl. Phys. A. 1-5 (2014)
7. L. Chen, Y. Zhou, C. Lei, Z. M. Zhou, W. Ding, Journal of Physics D: Applied Physics. 42(14), 145005 (2009)
8. G. V. Kurlyandskaya, M. L. Sanchez, B. Hernando, et al., Appl. Phys. Lett. 82, 3053 (2003).
9. R. L. Sommer, C. L. Chine, J Appl Phys. 79, 5139 (1996).

MATEC Web of Conferences **88**, 01004 (2017) DOI: 10.1051/matecconf/20178801004
CBNCM 2016

Luminescent Characteristics of Blue Phosphor $CaAl_2Si_2O_8:Eu^{2+}$ in White LED Applications by Solid State Method

Joo Hyun Lee[1], Sung Jun Park[1] and Hyun Kyoung Yang[1,a]

[1]*Department of LED Convergence Engineering, Pukyong National University, 48547, Busan, Republic of Korea*

Abstract. Using the solid state reaction method, we synthesize the $Ca_{0.88}Al_2Si_2O_8:Eu^{2+}_{0.12}$ phosphor powders at different sintering temperatures ranging from 1100 to 1500 °C. From X-Ray diffraction patterns, we confirmed that all the samples acquired mixed $CaAl_2Si_2O_8$ (triclinic), $CaAlO_4$ (Monoclinic) and SiO_8 (Hexagonal) phases. For the $Ca_{0.88}Al_2Si_2O_8:Eu^{2+}_{0.12}$ phosphors, the excitation spectra exhibited an unresolved broad band in the range from 200 to 420 nm and maximum intensity at 350 nm. The emission spectra of $CaAl_2Si_2O_8:Eu^{2+}$ show broad band ranging from 400 to 560 nm and maximum wavelength is found to be at around 440nm, which can be ascribed to the electric-dipole-allowed $4f^65d^1$ - $^8S_{7/2}(4f^7)$ transition of the Eu^{2+} ions. The Commission International De I-Eclairage 1931 (CIE 1931) chromaticity coordinates for $Ca_{0.88}Al_2Si_2O_8:Eu^{2+}_{0.12}$ phosphors are calculated and represented at (0.168, 0.144).

1 Introduction

Light emitting diodes (LED) are the promising candidates for next generation solid lighting material owing to their lower energy consumption, high brightness, environmentally-friendliness, fast switching, small in size, longer lifetime and high efficiency, which starts to approach the theoretical limits[1-2]. White light emitting diodes (WLEDs) are usually fabricated with blue InGaN LED chips and yellow phosphors such as Ce^{3+} doped yttrium aluminum garnet ($Y_3Al_5O_{12}:Ce^{3+}$), Eu^{2+} doped oxynitrides, and Eu^{2+}/Ce^{3+} doped silicate based materials[3-6].

Eu^{2+} have been widely used for activators in phosphor materials [7-8]. Much attention has been paid towards the luminescent properties of Eu^{2+} ions activated compounds and the reduction processes of Eu^{3+} to Eu^{2+} in phosphor preparations. The Eu^{2+} ion shows broad emission bands excited by ultra-violet (UV) light (320 - 400 nm), arising from interconfigurational $4f^65d^1$ (excitation state) – $4f^7$ (ground state) allowed transition [9-10]. This transition is strongly dependent on the crystal fields of the host lattices since 5d orbitals are more sensitive to the ligand field. Accordingly, the Eu^{2+} becomes a very useful activator in phosphors for applications in displays, lamps and luminescent paintings [11]. In general, the reduction of ionic state from Eu^{3+} to Eu^{2+} needs sintering process in reduce atmosphere

[a] Corresponding author : hkyang@pknu.ac.kr

such as H_2/N_2 or carbon. If the reduction of Eu^{3+} to Eu^{2+} can be realized in air condition, it would highly reduce the cost and increase the safety in preparing the Eu^{2+}-activated phosphor materials[12-14].

Aluminum silicate is well known as a long-persistent phosphor host with high chemical stability and water resistance. The alkaline-earth materials of $MAl_2Si_2O_8$ (M = Ca, Sr, Ba) have drawn much attention for their good luminescence properties [15-17]. These compounds exhibit strong $4f^65d^1 - 4f^7$ luminescence on doping Eu^{2+}. A series of solid solutions containing isostructural members with slightly different size cations as hosts, such as $MAl_2Si_2O_8$ (M = Ca, Sr, Ba) would be an ideal system to study the effect of crystal chemical variation on the luminescence of Eu^{2+} ions, and may offer possibilities for commercial application [18, 19].

In this paper, study of the solid state reaction, sintering temperature dependent photoluminescence characteristics of $CaAl_2Si_2O_8:Eu^{2+}$, is reported. Also, lifetime and CIE chromaticity coordinates of $CaAl_2Si_2O_8:Eu^{2+}$ have been be discussed with PL spectra behaviour.

2 Experiment and measurement

The powder of the $Ca_{0.88}Al_2Si_2O_8:Eu^{2+}{}_{0.12}$ phosphor was synthesized using a solid state reaction method. At first, $CaAl_2Si_2O_8:Eu^{2+}$ precursor powder was prepared by taking stoichiometric molar ratio of calcium carbonate ($CaCO_3$, 99.95%, Aldrich), silicon dioxide (SiO_2, 99.5 %, Aldrich), aluminium oxide (Al_2O_3, 99.9%, Aldrich), europium oxide (Eu_2O_3, 99.99 %, Aldrich). All regents were taken without any further purification and mixture, then powders were grinded and mixed by the planetary ball-milling method. Powders have been sintered at 1100 - 1500 °C. Here we have increase temperature from room temperature to given sintered temperature with 4 °C per a minute, then all samples are keep in sintered temperature for 4 hour in carbon atmosphere, respectively.

The sintered phosphors were ground using an alumina mortar. The thermal properties of the precursors were examined by a thermogravimetric-differential thermal analysis (TG-DTA) at a heating rate of 4 °C/min. And the crystal structure of the phosphor was obtained by X-Ray diffractometer (PHILIPS X'Pert-MPD) with CuKα = 1.546 Å radiation at 40 kV and 30 mA. The X-ray diffraction patterns show the intensities of the diffraction peaks as a function of the detecting angle 2θ. The interplaner spacing (d), matching peaks our samples and reference for JCDPS data, is calculated using the software X'pert High Score Plus. Miller indices were then assigned to the reflection peaks in the diffraction patterns by comparing the data obtained from the samples. Both the photoluminescence and lifetime measurements of emission bands of the phosphors were recorded on the luminescence spectrophotometer (Photon Technology International (PTI)) with a Xe-arc lamp (60 W) as an excitation source.

3 Results and discussions

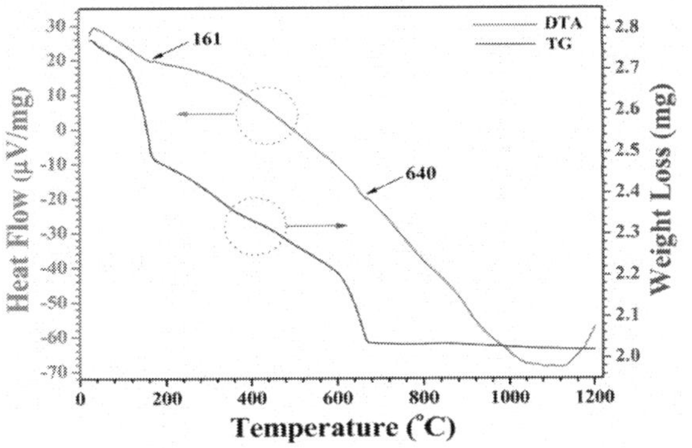

Figure 1. Thermogravimetric-differential thermal analysis curves of $CaAl_2Si_2O_8$ precursor in air

DOI: 10.1051/matecconf/20178801004

Figure 1 shows the TG-DTA measurements of $CaAl_2Si_2O_8$ precursor, which were carried out to monitor the decomposition process in the host lattice. The endothermic peaks and weight loss were observed until 161 °C. The weight loss is originated from the evaporation of water molecules in the precursor samples. In the heat absorbing section between 161 and 640 °C, one can see the weight loss again. It means that the absorbing section arise from the weight loss due to the decomposition of the remaining evaporation of water molecules and organic material. After 660 °C, endothermic behaviour without weight loss is confirmed and it means that remaining materials participate in the synthesis.

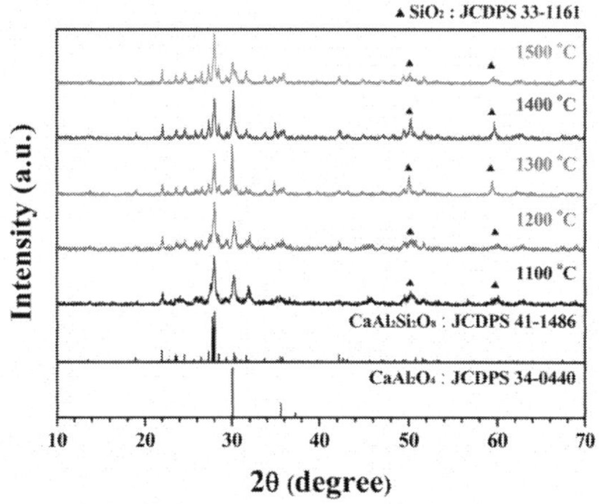

Figure 3. XRD patterns of $Ca_{0.88}Al_2Si_2O_8:Eu^{3+}_{0.12}$ phosphors at different sintering temperatures

The XRD pattern of $Ca_{0.88}Al_2Si_2O_8:Eu^{2+}_{0.12}$ phosphor synthesized at 1100 - 1500 °C for 4hour under reduce atmosphere are shown in Figure 2. The main peaks of phosphors from XRD pattern corresponding to $CaAl_2Si_2O_8$ (JCDPS 41-1486, triclinic), $CaAl_2O_4$ (JCDPS 34-0440, Monoclinic) and SiO_2 (JCDPS 33-1161, Hexagonal) phase. Samples were crystallized into a triclinic phase with a space group of P-1. The calculated lattice constants sintered at 1300 °C were a = 8.1430 Å, b = 12.8789 Å, and c = 14.1828 Å , which are very close to the corresponding values (a = 8.1756 Å, b = 12.8720 Å, c = 14.1827 Å) in the standard JCPDS 41-1486. We have observed the increasing the intensity of the $CaAlO_4$ until 1300 °C, then the intensity gradually decreasing, and opt-out silicone oxide in the reaction is confirmed by increasing intensity of SiO_2 marked as triangle body in Fig 2.

Figure 2. XRD patterns of $Ca_{0.88}Al_2Si_2O_8:Eu^{3+}_{0.12}$ phosphors as a function of sintering temeperature

FE-SEM images in Figure 3(a) - (e) of $Ca_{0.88}Al_2Si_2O_8:Eu^{2+}{}_{0.12}$ phosphor represent the particle shape and surface at each sintering temperature. The particles shows increasing irregular shape with increasing the sintering temperature. Also, the average particle and the crystallite sizes are increases depending on sintering temperature. The average crystallite sizes are displayed in Fig 5. Interestingly, we have observed the significantly increased average crystallite sizes until 1300 °C. After 1300 °C, the average crystallite size is almost independent on sintering temperature. Note that the particle size is increased with temperature. This can be explained by the Ostwald ripening process [20]. As shown in Figure 3(f), EDS was used to examine the elemental composition of $Ca_{0.88}Al_2Si_2O_8:Eu^{3+}{}_{0.12}$ powder sintered at 1300 °C. The EDS spectrum of the $Ca_{0.88}Al_2Si_2O_8:Eu^{3+}{}_{0.12}$ powder shows the presence of Calcium (Ca), Aluminum (Al), Silicone (Si) and Europium (Eu), but other components are not detected.

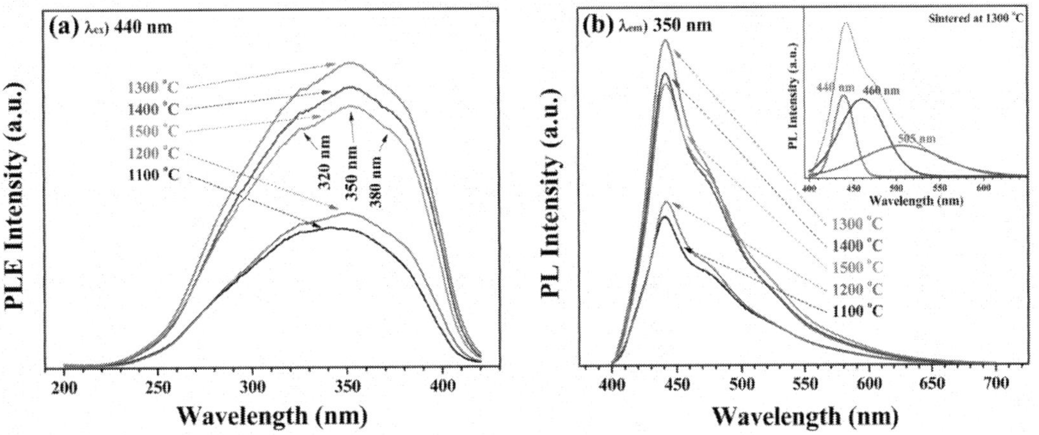

Figure 4. (a) PLE and (b) PL spectra of $Ca_{0.88}Al_2Si_2O_8:Eu^{2+}{}_{0.12}$ phosphors as a function of sintering temperature

Figure 4 shows the measured photoluminescence (a) excitation (PLE) and (b) emission (PL) spectra of $Ca_{0.88}Al_2Si_2O_8:Eu^{2+}{}_{0.12}$ at different sintering temperatures. The PLE spectrum monitored by the emission at 440nm exhibited a broad band within the range from 200 to 420 nm, which can be assigned to the transitions from the ground state $^8S_{7/2}(4f^7)$ to the crystal-field-split components of the $4f^65d^1$ configuration. The excitation band represent a triple overlapping peaks cantered at 320, 350 and 380 nm. The 320 nm (t_{2g}) and 380 nm (e_g) are due to the result of 5d orbital separation caused by crystal field splitting peaks, and 350nm is excitation by $CaAl_2O_4$ [21-22]. The $Ca_{0.88}Al_2Si_2O_8:Eu^{3+}{}_{0.12}$ phosphor displayed blue luminescence excited by near-UV light. The emission spectrum of $Ca_{0.88}Al_2Si_2O_8:Eu^{2+}{}_{0.12}$ have a broad band from 400 to 670 nm with a maximum wavelength of about 450 nm, which can be ascribed to the electric dipole-allowed $4f^65d^1$-$8S_{7/2}(4f^7)$ transition of the Eu^{2+} ions. PL spectra indicate wide emission region extending to green region. It is obtained triple multi emission band consisted with wave length of 440, 460, and 505 nm by Gaussian-fitting method as shown in figure 4(b). The transition from $5D^1$ to be ground states can be split by crystal field. Actually, the emission bands are possible due to the allowed character of the 5d → 4f transitions, and also to the existence of 460 nm of alkaline earth sites available for the activator. Although Eu^{2+} can occupy into the interstitial site in SiO_2 crystal [23]. Therefore, the overlapped 460nm and 505nm are due to different Eu^{2+} site in $CaAl_2Si_2O_8$ and this make different environment to Eu^{2+}. Also, the other emission peak (440 nm) is caused by $CaAl_2O_4$ [24]. From the inset of Figure 4(b), one can see the strongest peak in the blue region described by the coupling of 440 nm and 460 nm. In the emission spectrum, a third contribution in the yellow range around 560 nm is clearly visible. Phosphor converted LEDs (pc-LEDs), based on a blue $In_{1-x}Ga_xN$ diode and a yellow phosphor which is kind of $Y_3Al_5O_{12}:Ce^{3+}$[3-6], suffered from a low color rendering due to the lack of green and red light in the spectrum and of inherently high correlated color temperatures (CCT). To solve this problem consists by using a cyan phosphor with a small Stokes shift and uses a violet or near UV excited diode in

DOI: 10.1051/matecconf/20178801004

combination with an additional blue phosphor. Therefore, it is completely explained that $Ca_{0.88}Al_2Si_2O_8:Eu^{2+}{}_{0.12}$ phosphors have good color rendering.

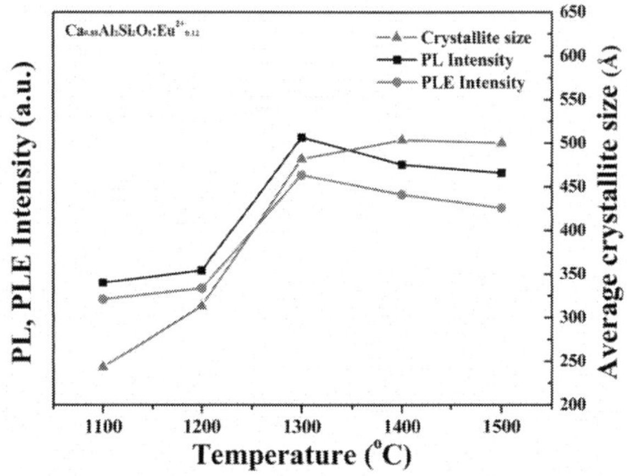

Figure 5. Comparison of PL, PLE intensity and average crystallite size of $Ca_{0.88}Al_2Si_2O_8:Eu^{2+}{}_{0.12}$ phosphors at the sintering temperatures ranging from 1100 - 1500 °C

Figure 5 shows comparison of PL, PLE intensity and average crystallite size of $Ca_{0.88}Al_2Si_2O_8:Eu^{2+}{}_{0.12}$ phosphors at different sintering temperatures from 1100 - 1500 °C. We have calculated average crystallite size using the Scherrer equation expressed as $D = k\lambda/\beta\cos\theta$, where D is the average grain size, k (0.9) is a shape factor, λ is the Cu-Kα wavelength (1.5406 Å) of X-ray, β is the full width at half maximum (FWHM) and θ is the diffraction angle of an observed peak, respectively. PLE and PL intensity is well matched with the average crystallite size. PLE and PL intensity is to reduce, and average crystallite size is increased above 1400 °C. PL and PLE intensity are affected by $CaAl_2Si_2O_8$ and $CaAl_2O_4$. PL and PLE spectra of $CaAl_2Si_2O_8$ and $CaAl_2O_4$ have a similar region. This result is in good agreement with XRD patterns. Therefore, the maximum PL and PLE intensity was confirmed at 1300 °C for both $CaAl_2Si_2O_8$ and $CaAl_2O_4$.

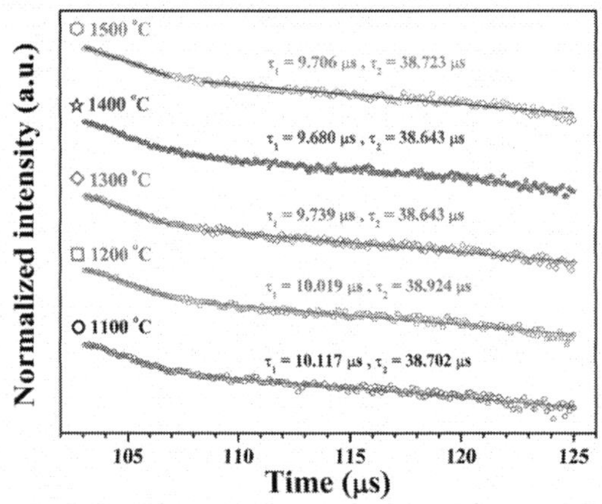

Figure 6. Luminescence decay curves at room temperature of $Ca_{0.88}Al_2Si_2O_8:Eu^{3+}{}_{0.12}$ phosphors as a function of sintering temperature

The lifetime decay curves from $4f^65D^1 \rightarrow 4F^7$ transition of Eu^{2+} are shown in Figure 6. Decay curve of the all samples can be fitted by a double-exponential function,

$$I(t) = I_1exp(-t/\tau_1) + I_2exp(-t/\tau_2) \rightarrow lnI(t) = ln[I_1exp(-t/\tau_1) + I_2exp(-t/\tau_2)] + C \qquad (1)$$

where I is a luminescence intensity at time, I_1, I_2 and C are constants, τ_1 and τ_2 are decay times. The lifetime for the emission of $Ca_{0.88}Al_2Si_2O_8:Eu^{2+}{}_{0.12}$ sintered at 1300 °C sample is τ_1 = 9.739 μs and τ_2 = 38.643 μs. This decay properties are independent on sintered temperature or defects on $Ca_{0.88}Al_2Si_2O_8:Eu^{2+}{}_{0.12}$ powders such as $CaAl_2O_4$ and SiO_2. It is explained that $CaAl_2O_4:Eu^{2+}$ and $CaAl_2Si_2O_8:Eu^{2+}$ phosphors have same emitting mechanism ($4f^65D^1 \rightarrow 4F^7$ transition of Eu^{2+}) and similar emission region.

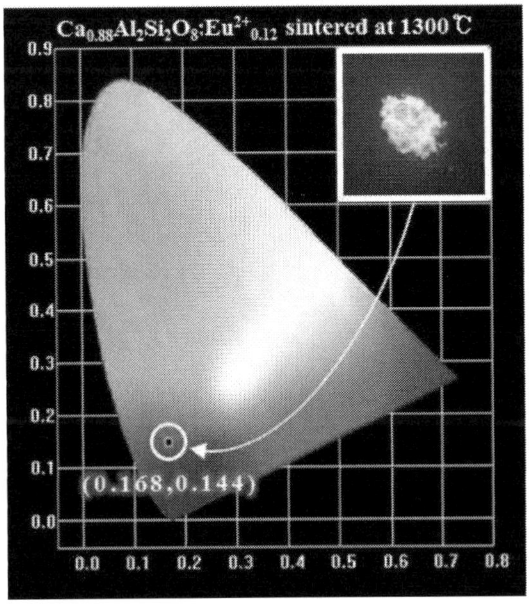

Figure 7. CIE 1931 color coordinates of $Ca_{0.88}Al_2Si_2O_8:Eu^{2+}{}_{0.12}$ phosphor sintered at 1300 °C. The inset picture shows the photographs of blue light emission by near-UV light

Table 1. CIE 1931 color coordinates of $Ca_{0.88}Al_2Si_2O_8:Eu^{2+}{}_{0.12}$ phosphor sintered at 1300 °C

Sintering temperature (°C)	X	Y
1100	0.176	0.166
1200	0.172	0.158
1300	0.168	0.144
1400	0.171	0.156
1500	0.175	0.172

CIE 1931 chromaticity coordinates of $Ca_{0.88}Al_2Si_2O_8:Eu^{2+}{}_{0.12}$ phosphor for different sintering temperature displayed in Table 1. In addition, CIE 1931 color coordinate of $Ca_{0.88}Al_2Si_2O_8:Eu^{2+}{}_{0.12}$ phosphor was presented in Figure 7. The CIE coordinate of $Ca_{0.88}Al_2Si_2O_8:Eu^{2+}{}_{0.12}$ phosphor sintered at 1300 °C located at blue region (0.168, 0.144), and CIE coordinate of other samples are almost similar values.

4 Conclusions

The $Ca_{0.88}Al_2Si_2O_8:Eu^{2+}{}_{0.12}$ phosphors were prepared by a solid state reaction method. The sintering temperature influences the structural properties, which were confirmed by analysing the XRD patterns, SEM images and luminescence properties. XRD patterns indicate that phosphors have mixed phases of both $CaAl_2Si_2O_8$ (JCDPS 41-1486 triclinic) and $CaAl_2O_4$ (JCDPS 34-0440 Monoclinic). The Eu^{2+} ion was well occupied in monoclinic crystalline structure. Luminescence properties of

$Ca_{0.88}Al_2Si_2O_8:Eu^{2+}_{0.12}$ have PL (λ = 440 nm) and PLE (λ = 350 nm) maximum intensity at 1300 °C. PL spectra have emission region extending to green region. Hence, based on the short life time, CIE coordinates and wide blue region of the broad PL spectra, $Ca_{0.88}Al_2Si_2O_8:Eu^{2+}_{0.12}$ could be a potential blue color emitting candidate, which can be used to improve the quality of white.

Acknowledgement

This work was supported by the Human Resource Training Program for Regional Innovation and Creativity through the Ministry of Education and National Research Foundation of Korea (NRF-2014H1C1A1066586).

References

1. M. Zhang, J. Wang, Q. Zhang, W. Ding and Q. Su, Mater. Res. Bull. 42, 33 (2007)
2. Y. Shi, Z. Yang, W. Wang, G. Zhu and Y. Wang, Mater. Res. Bull. 46, 1148 (2011)
3. R. J. Xie, N. Hiroaski, M. Mitomom, K. Sakuma and N. Kimura, Appl. Phys. Lett. 89 241103 (2006)
4. H. S. Jang and D. Y. Jeon, Appl. Phys. Lett. 90 041906 (2007)
5. P. Li, L. Pang, Z. Wang, Z. Yang, Q. Guo and X. Li, J. Alloys Compd. 478, 813 (2009)
6. H. K. Yang, H. M. Noh and J. H. Jeong, Solid State Sciences. 27, 43 (2014)
7. Y. K. Kim, D. H. Cho, Y. K. Jeong and M. K. Nah, Mater. Res. Bull. 45, 905 (2010)
8. L. Guan, C. Liu, X. Li, G. Jia, Q. Guo, Z. Yang and G. Fu, Mater. Res. Bull. 46, 1496 (2011)
9. S. H. M. Poort, W. P. Blokpoel and G. Blasse, Chem. Mater. 7, 1547 (1995)
10. C. M. Zhang, J . Yang, C. K. Lin, C. X. Lin and J. Lin, J. Solid State Chem. 182, 1673 (2009)
11. M. Y. Peng, Z. W. Pei, G. Y. Hong and Q. Su, J. Mater. Chem. 13, 1202 (2003)
12. Z. H. Lian, J. Wang, Y. H. Lv, S. B. Wnag and Q. Su, J. Alloys compd. 430, 257 (2007)
13. I. Ta¨le, P. Ku¨lis and V. Kronghauz, J. Lumin. 20, 343 (1979)
14. M. Y. Peng, Z. W. Pei, G. Y. Hong and Q. Su, Chem. Phys. Lett. 371, 1 (2003)
15. Y. J. Hua, S. Q. Xu, D. G. Deng, S. L. Zhao, H. P. Wang and L. H. Huang, Optics Commun. 284, 27 (2011)
16. B. L. Wang, L. Z. Sun, H. D. Ju, S. L. Zhao, D. G. Deng, H. P. Wang and S. Q. Xu, J. Sol–Gel Sci. Technol. 50, 368 (2009)
17. K. Y. Jung and J. H. Kim, J. Lumin. 128, 2004 (2008)\
18. K. R. Laud, E. F. Gibbons, T. Y. Tien and H. L. Stadler, J. Electrochem. Soc. 118, 918 (1971)
19. Z. Wu, J. Liu and M. Gong, Chem. Phys. Lett, 466, 88 (2008)
20. D. Alloyeau, C. Ricolleau, C. Mottet, T. Oikawa, C. Langlois, Y. Le Bouar, N. Braidy and A. Loiseau, Nature Mater. 8, 940 (2009)
21. F. Clabau, A. Garcia, P. Bonville, D. Gonbeau, T. L. Mercier, P. Deniard and S. Jobic, J. Solid State Chem. 181, 1456 (2008)
22. T. Aitasaloa, J. Hölsäa, H. Jungnerd, M. Lastusaaria and J. Niittykoski, J. Alloys Compounds, 341, 76 (2002)
23. B. S. Kim , Y. H. Song, D. S. Jo, T. Abe, K. Senthil, T. Masaki, K. Toda and D. H. Yoon, Materials Research Bulletin, 47, 3015 (2012)
24. F. C. Palilla, A. K. Levine and M. R. Tomkus, J. Electrochem. Soc. 115, 643 (1968)

The Preparation of VO_2(B) Cathode Material for Lithium-ion Battery with High Capacity and Good Cycling Performance

Zhongliang Hou[1], Zhengguang Zou[1,a], Zhendong Wan[1] and Shichang Han[1]

[1]*Key Laboratory of Non-ferrous Materials and New Processing Technology of Ministry of Education, College of Material Science and Engineering, Guilin University of Technology, Guilin 541004,China*

Abstract. Fiber-like VO_2(B) was successfully synthesized by using V_2O_5 and ethanol as reactants via a magnetic stirring solvothermal process. The stirring rates significantly affected the phase, morphology and the cycling performance of as-synthesized products. When the stirring rate was 867 rpm, the fiber-like particles were 3-5 μm long and 50-100 nm wide, and showed better dispersion than the sample of VO-0, the electrochemical performance test demonstrated that the initial discharge capacity of VO-867 was 223 mAh/g, and maintained 186 mAh/g after cycling for 50 times, the retention rate of the capacity was 83.4%, which showed best cycling property of all samples.

1 Introduction

Nowadays, the deteriorating of the ecological environment and the depleting of the traditional energy made the development of human society face enormous challenges, searching for an alternative and efficient green energy, has become an imperative things for the sustainable development of human society[1-3]. The advantage of high energy density, long life, no memory effect, safety and environmental made the Lithium-ion batteries be the best energy storage equipment[4-7].

As we know that the successful application of lithium-ion battery technology was highly depended on the used of the electrode material, so how to choose the electrode material would in turn limits the safety, energy density and the power density of the lithium-ion battery system. Compared with the conventional lithium-ion battery cathode material ($LiCoO_2$[4], $LiFePO_4$[8]), vanadium oxides (V_2O_5, V_6O_{13}, VO_2, etc.) with the advantage of low cost, high specific capacity, high security and easy of preparation have been widely researched in recent years [9,10]. Among which, VO_2(B) has been one of the most promising cathode materials due to its open framework structure which permitted the lithium ions rapidly intercalation/deintercalation between the layers, the special structure gave it a maxium reversible capacity of 320 mAh/g in the 4-1 V voltage range[11-14]. However, the defect of low electronic conductivity and poor stable circulation severely affected its practical application. In recent years, researchers have taken a series of effective measures to overcome its shortcomings, all kinds of morphologies of VO_2(B) have been prepared, such as nanorods[15,16], nanoribbons[17,18], nanotubes[19,20], and nanowires[21], however, these nanostructure materials displayed low discharge capacity and poor cycling properties[22], The fading of the capacity was probably owing to the high surface energies and large specific surface areas, which leading to the easier aggregation of the sample during process of charging and discharging, thus resulting in the increases of the charge transfer

[a] Corresponding author : zouzgglut@163.com

resistance[23]. Therefore, how to improve the cycling properties of VO_2(B) was still a great challenge for researchers and material scientists.

Yuxin Tang and co-workers[24] proposed a stirring hydrothermal method and successfully synthesized the elongated titanate nanotubes with length up to tens of micrometers, meanwhile, this method effectively prevented agglomeration of the sample, and provided a pathway for electron transport, thus improved the electrochemical performance greatly. In this work, Fiber-like VO_2(B) was successfully synthesized by using V_2O_5 and ethanol as reactants via a magnetic stirring solvothermal process. Afterwards, we discussed the influence of the mangnetic stirring rate on the phase, morphologies and electrochemical properties of the samples.

2 Experimental

2.1 synthesis

All the reagents were analytical grade and used without any further purification. In a typical synthesis, 0.4 g V_2O_5 was dispersed into the solution of 20 mL deionized water and 30 mL anhydrous ethanol, after stirring for several minutes, the uniformly yellow suspension was obtained and transferred into a 100 mL teflon lined autoclave. The autoclave was placed into the oil bath and heated to 190 °C for 24 h with magnetic stirring, and then cooled to room temperature naturally, the products were washed with deionized water for 3 times and then dried in a vacuum freeze-dryer for 24 h. In order to investigate the relation between stirring rate and morphology, the stirring rate was regulated for 0 rpm, 867 rpm and 1730 rpm, and the products were marked as VO-0, VO-867 and VO-1730, respectively.

2.2 Characterization

The X-ray powder diffractometer (XRD) with a Cu Kα radiation source (λ = 0.154 nm, the scanning rate of 6 °/min) was used to analyze the phase of the products. The Hitachi S-4800 field emission scanning electron microscopy (FESEM) was carried out to observe the morphology of the products.

2.3 Electrochemical performance measurements

The positive electrode was prepared by mixing active material, acetylene black carbon power and polyvinylidene fluoride (PVDF) binder in a weight ratio of 7/2/1, grinding the mixture with some N-methyl-2pyrrolidone (NMP), then the slurry was coated equably on Al foil. The Al foil was dried in vacuum at 90 °C for 12 h.The 1mol/L LiPF6 mix with ethylene carbonate (EC), dimethyl carbonate (DMC) and diethyl carbonate (DEC) in a volume ratio of 2/2/1, then assembled the cells in the glove box. Charge-discharge tests were performed by a BTS Cycler (Shenzhen, China), the Cyclic voltammetry (CV) measurement and electrochemical impedance spectroscopy(EIS) measurement were characterized by a Zahner (Germany) IM6ex electrochemical workstation.

3 Results and discussion

3.1 XRD analysis

Figure 1 showed the XRD patterns of the samples with different stirring rates, as shown in figure 1, the peaks of the samples were basically agreement with standard diffraction peaks of VO_2(B) (JCPDS no. 81-2392). However, the characteristic diffraction peaks of V_3O_7 and V_2O_5 have been found near 14° and 26° respectively, which may be attributed to the reaction was not completely.

DOI: 10.1051/matecconf/20178801005

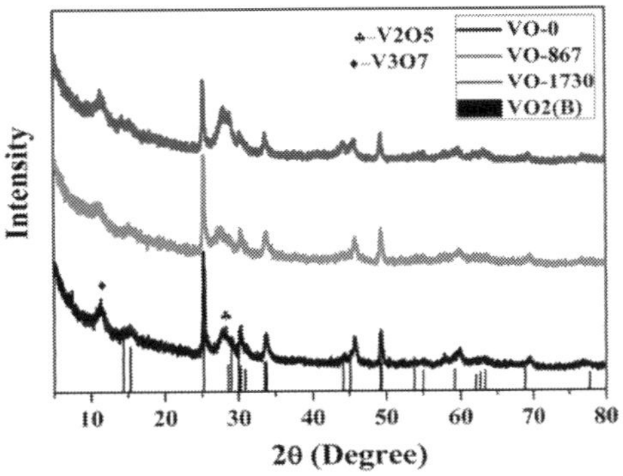

Figure 1. XRD patterns of the samples with different stirring rates.

3.2 SEM analysis

Figure 2 showed the SEM images of the samples with different stirring rates of VO-0(a,b), VO-867(c,d) and VO-1730(e,f), respectively. It can be seen from figure 2(a,b) that the sample with no stirring speed was composed of fiber-like particles, the fibrous were 1-2μ m long and 100-300nm wide, what is more, they were easy to aggregation. With the stirring rate increasing, the length of the fibrous increased greatly, meanwhile, the width of them decreased significantly, when the stirring speed increased to 867 rpm, the fiber-like particles can approach to 3-5μ m long and 50-100 nm wide (figure c and figure d), however, when the stirring speed increased to 1730 rpm (figure e and figure f), the length of the fibrous were shorter than that of the sample of VO-867, indicating that too high of the speed would destroy the structure of the sample, which would further influence the electrochemical performance.

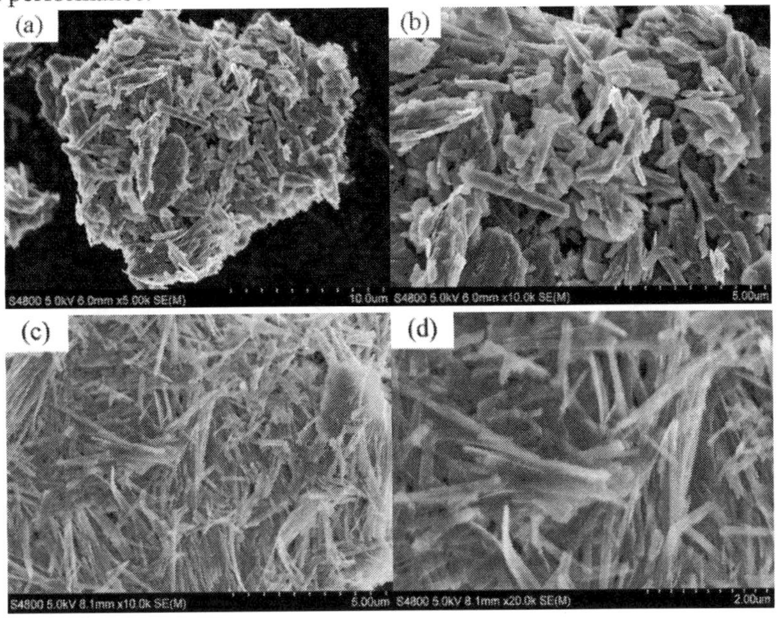

 DOI: 10.1051/matecconf/20178801005

Figure 2. SEM images of the samples with different stirring rates of VO-0(a,b), VO-867(c,d) and VO-1730(e,f).

3.3 Electrochemical properties

Figure 3 showed the Nyquist plots of the samples, the Nyquist plots were measured by EIS (electrochemical impedance spectra) after cycling for 3 times, It can be seen from figure 3 that the charge transfer impedance (Rct) value of VO-0, VO-867 and VO-1730 in the high frequency region's semicircles were corresponding to 908 Ω, 551.5 Ω and 762.4 Ω, obviously the Rct value of the samples with stirring speed were smaller than that of the sample with no stirring rate, among which, the Rct value of VO-867 was found to be the smallest, indicating that the sample of fiber-like can be better contact with the conductive material in battery, which makes uniform voltage distribution and smaller resistance during the process of lithium ion intercalating to the surface of the electrode.

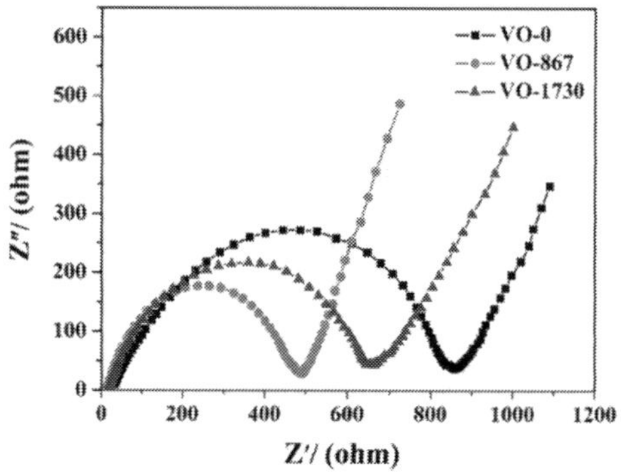

Figure 3. the Nyquist plots of the samples with different stirring rates.

Figure 4 displayed the cycling performance of the samples with different stirring rates of VO-0, VO-867 and VO-1730 at the current density of 32.4 mA/g (0.1 C), it can be seen from figure 6 that the discharge specific capacities were 308 mAh/g, 223 mAh/g and 291 mAh/g after the first cycle, corresponding to the samples of VO-0, VO-867 and VO-1730, respectively, and maintained 177 mAh/g, 186 mAh/g and 144 mAh/g after 50 cycles, the retention rates of the capacities were 57.4%, 83.4% and 49.5%, respectively. Obviously, the sample of VO-867 had the best cycling performance, which was attributed to its special fibrous structure that composed of nanobelts, these nanobelts were more longer and thinner than the other two samples of VO-0 and VO-1730, meanwhile, they were stacking and intertwined together, which permitted the faster transmission of lithium-ion and reduce the damage to the structure, thus greatly improved the cycling performance of the sample.

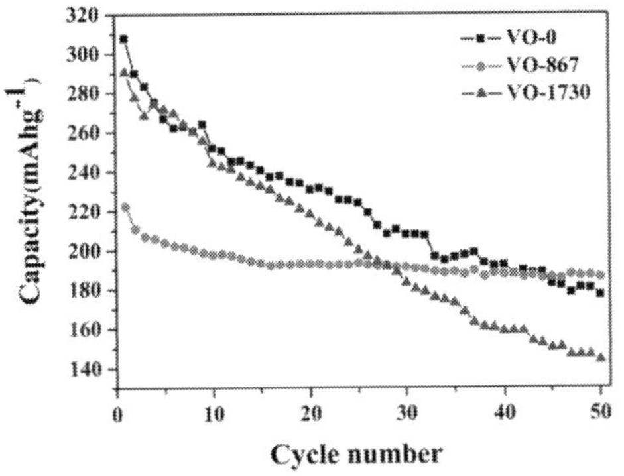

Figure 4. the cycling performance of the samples.

Figure 5 showed the cyclic voltammetry curves of VO-0, VO-867 and VO-1730 with the voltage range of 1.5-3.5 V at the scanning speed of 0.2 mV/s after cycling for 3 times. A pair of oxidation and reduction peaks could be observed in the figure, which represented the extraction and insertion of Li^+ during the electrochemical processes. With the stirring rate increasing, the reduction peaks shift to higher voltage and the oxidation peaks shift to lower voltage, of all the samples, the area of the curve in the sample of VO-0 was the biggest, which represented the sample of VO-0 had the biggest capacity, In addition, the peaks of VO-867 were highest symmetric, indicating that its cycling properties was the best. Which was consistent with the result of the cycling performance.

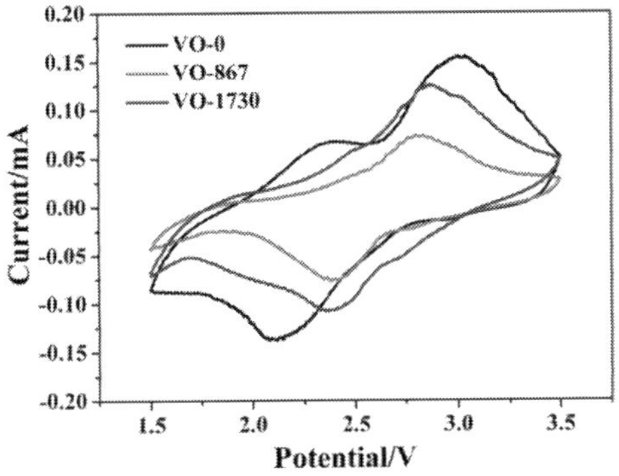

Figure 5. The cyclic voltammetry(CV) curve of the samples.

4 Conclusion

VO_2(B) with fibrous structure has been successfully synthesized via a solvothermal method with magnetic stirring in a oil bath. The morphology of the sample was controlled by the stirring rate. With stirring rate increasing, the length of the fibrous increased and the width of them decreased significantly, meanwhile, they were better dispersion, when the stirring rate increased to 867 rpm, the fibrous can approach to 3-5μ m long and 50-100 nm wide, electrochemical performance test demonstrated that the initial discharge capacity of VO-867 was 223 mAh/g, and maintained 186 mAh/g after cycling for 50 times, the retention rate of the capacity was 83.4%, which showed best cycling property of all samples.

References

1. Xiaoqing Zhang, Zhimin Zhao, Coal Techno, **29**, 9-10(2010)
2. Ting Wan, Daobin Mu, Huan Xue, Shi Chen, Mater. Rev, **24**, 117-120(2010)

3. Ping Du, Junkui Gao, Chin. J. Power Sources, **4**, 409-412 (2010)
4. Kisuk Kang, Ying Shirley Meng, Julien Breger, Clare P. Grey, Gerbrand Ceder, Sci, **311**, 977-980(2006)
5. Hao Bin Wu, Anqiang Pan, Huey Hoon Hng, Xiong Wen (David) Lou, Adv. Funct. Mater, **23**, 5669-5674(2013)
6. Anqiang Pan, Hao Bin Wu, Lei Zhang, Xiong Wen (David) Lou, Energy Environ. Sci, **6**, 1476-1479(2013)
7. Zhongli Wang, Dan Xu, Limin Wang, Xinbo Zhang, ChemPlusChem, **77**, 124-128(2012)
8. Jiajun Wang, Xueliang Sun, Energy Environ. Sci, 5, 5163-5185(2012)
9. C. Leger, S. Bach, J-P. Pereira-Ramos, J. Solid State Electrochem, **11**, 71-76(2005)
10. W. J. Macklin, R. J. Neat, S. S. Sandhu, Electrochim. Acta, **37**, 1715-1720(1992)
11. D. W. Murphy, P. A. Christian, F. J. DiSalvo, J. V. Waszczak, Sci, **205**, 651-656(1979)
12. J. Glanz, Sci, **264**, 1084(1994)
13. Natasha A. Chernova, Megan Roppolo, Anne C. Dillon, M. Stanley Whittingham, J. Mater. Chem, **19**, 2526-2552(2009)
14. Emmanuel Baudrin, Guillaume Sudant, Dominique Larcher, B. Dunn, Jean-Marie Tarascon, Chem Mater, **18**, 4369-4374(2006)
15. Zhanjun Chen, Shaokang Gao, Lilong Jiang, Mingdeng wei, Kemei Wei, Mater. Chem. Phys, **121**, 254-258(2010)
16. Ch. V. Subba Reddy, E. H. Walker Jr., S. A. Wicker Sr., Quinton L. Williams, Rajamohan R. Kalluru, Curr. Appl. Phys, **9**, 1195-1198(2009)
17. Lijuan Mao, Chunyan Liu, Mater. Res. Bull, **43**, 1384-1392(2008)
18. Guicun Li, Shuping Pang, Li Jiang, Zhiyan Guo, Zhikun Zhang, J. Phys. Chem. B, **110**, 9383-9386(2006)
19. M. Niederberger, H.-J. Muhr, F. Krumeich, F. Bieri, D. Gunther, R. Nesper, Chem. Mater, **12**, 1995-2000(2000)
20. F. Krumeich, H.-J. Muhr, M. Niederberger, F. Bieri, B. Schnyder, R. Nesper, J. Am. Chem. Soc, **121**, 8324-8331(1999)
21. X.Y. Chen, X. Wang, Z.H. Wang, J. X. Wan, J. W. Liu, Nanotechnol, **15**, 1685-1687(2004)
22. S. D. Zhang, Y. M. Li, C. Z. Wu, F. Zheng, Y. Xie, J. Phys. Chem. C, **113**, 15058-15067(2009)
23. Haimei Liu, Yonggang Wang, Kaixue Wang, Eiji Hosono, Haoshen Zhou, J. Mater. Chem, **19**, 2835-2840(2009)
24. Yuxin Tang, Yanyan Zhang, Jiyang Deng, Jiaqi Wei, Hong Le Tam, Bevita Kallupalathinkal Chandran, Zhili Dong, Zhong Chen, Xiaodong Chen, Adv. Mater, **26**, 6111-6118(2014)

 DOI: 10.1051/matecconf/20178801006

Hydrothermal Synthesis of Al/Cr-doped V_6O_{13} as Cathode Material for Lithium-ion Battery

Qi Yuan[1] and Zhengguang Zou[1,a]

[1]*College of Material Science and Engineering, Guilin University of Technology, 541004 Guilin, China*

Abstract. Pure V_6O_{13} and Al/Cr-doped V_6O_{13} were synthesized via a hydrothermal route using $C_2H_2O_4 \cdot 2H_2O$, V_2O_5, $Al(NO_3)_3 \cdot 9H_2O$ and $Cr(NO_3)_3 \cdot 9H_2O$ as raw materials. The products were characterized by XRD, SEM, EDS. Doping proven to be an effective method to improve the samples discharge specific capacity and cycle performance. Doping samples electrochemical performance were better than pure V_6O_{13}, the initial discharge specific capacity of sample 0.02 and 0.06 were 311mAh/g and 337mAh/g larger than pure V_6O_{13} sample (241 mAh/g). The capacity retention of samples 0.00, 0.02, 0.06 was 32.0%, 44.69%, 28.78% after 100 cycles, respectively. The increased electrochemical performance originated from the enhanced of electrical conductivity and adhered together by stacking region in an regular arrangement with every unit.

1 Introduction

Cathode materials capacity was the key of the whole specific capacity for lithium ion battery. Vanadium atoms which had various oxidation states (+2, +3, +4, +5), including different kinds of single-valence and mix-valence oxides for example VO_2, V_6O_{13}, V_3O_7 and V_2O_5. These compounds had partially filled d-orbitals which mean that a particular electrochemical property [1]. Among them, V_6O_{13} become the candidate of high performance cathode materials for lithium ion batteries because it theoretical specific capacity was high (420mAh/g) and electronic Conductivity [2, 3].

The number of conduction electrons in V_6O_{13} crystal was limited so electrical conductivity fell quickly followed by lithium ion insertion that reduced the utilization coefficient of the active material of cathode materials in high content Li. Moreover, V_6O_{13} volume changed and poorly stability of crystal structures followed by lithium ion insertion-extraction lead to cycle performance was falling fast[4]. Therefore, the key of influence V_6O_{13} application was how to efficiently improve $Li_XV_6O_{13}$ electrical conductivity and cycle performance in high content Li. Practices shown that it without a good effect to add the material with high conductivity directly in V_6O_{13}. The V_6O_{13} cathode material in the process of lithium insertion and extraction research by J. Barker[5] shown that resistance change correspond to electrode material crystal lattice large change in the process of lithium insertion and extraction. In the process of lithium insertion research based J. O. Thomas [6,7] shown that as lithium ion insertion V_6O_{13} layers to form $Li_xV_6O_{13}$, lithium ion occupied tetragonal pyramid vacancy between dioctahedron and octahedron, so the key of improve $Li_xV_6O_{13}$ cycle performance was reduce the internal resistance from the interface of electrolyt and $Li_xV_6O_{13}$ to $Li_xV_6O_{13}$ crystal between dioctahedron and octahedron meanwhile keep stability of crystal structures in

[a] Corresponding author :zouzgglut@163.com

the process of lithium insertion as possible. J. P. Pereira-Ramos [8, 9] synthesized $Cr_{0.36}V_6O_{13.5}$ compounds through chromium ions doped with V_6O_{13} with initial discharge capacity 370 mAh/g and capacity loss less than 15% after 35 cycles.

In this paper, pure V_6O_{13} and Al/Cr-doped V_6O_{13} were synthesized via a completely aqueous solution based synthesis method. Synthesis of different amounts Al/Cr doped V_6O_{13} were studied. Moreover, electrochemical performance of the pure V_6O_{13} and Al/Cr-doped V_6O_{13} were determined by charge/discharge tests and cyclic voltammetry. Sample designations with doping different amounts of Aluminum nitrate nonahydrate ($Al(NO_3)_3\cdot 9H_2O$) and Chromium trinitrate nonahydrate ($Cr(NO_3)_3\cdot 9H_2O$) as Table 1.

Table 1. Sample designations with doping different amount of Aluminum nitrate nonahydrate ($Al(NO_3)_3\cdot 9H_2O$) and Chromium trinitrate nonahydrate($Cr(NO_3)_3\cdot 9H_2$).

Sample designation	$Al(NO_3)_3\cdot 9H_2O/Cr(NO_3)_3\cdot 9H_2O$(g)
0.00	0.00/0.00
0.02	0.06/0.02
0.06	0.06/0.06

2 Experimental

2.1 Synthesis

All chemical reagents were analytical grade and used without any further purification. Al/Cr-doped V_6O_{13} were prepared according to the following procedure. 1.25g Oxalic acid dihydrate ($C_2H_2O_4\cdot 2H_2O$) and 0.4g Vanadium pentaoxide (V_2O_5) was dissolved in 20mL of deionized water. The mixed solution was kept under stirring at 80℃ in a water bath until a blue colored solution formed. The reaction is followed.

$$V_2O_5 + 3H_2C_2O_4 \rightarrow 2[(VO)(C_2O_4)](blue) + CO_2 + 3H_2O$$

Taked it out and the solution was cooled to room temperature naturally.The solution was filtered. Suitable amount of Aluminum nitrate nonahydrate ($Al(NO_3)_3\cdot 9H_2O$) and Chromium trinitrate nonahydrate ($Cr(NO_3)_3\cdot 9H_2O$) was dissolved in 15mL of deionized water, when it completely dissolves, pour it into as-synthesized Vanadyl oxalate ($VOC_2O_4\cdot 5H_2O$) solution, then 3ml Hydrogen peroxide 30%(H_2O_2) was added to mixed solution, solution comed up to lots of bubbles and a red solution was formed.The red mixed solution was transferred into a 100mL stainless steel autoclave after stirring it without comed up to bubbles. The autoclave was sealed and maintained at 160 ℃ for 24 h and then cooled to room temperature naturally. The supernatant liquid was discarded, suitable amount of deionized water was added to precipitate and the precipitate was separated by centrifugation (4000 rpm for 5 min). This process was repeated twice more after additional rinsing. Sample drying through freeze-drying process for 24h. After the dried sample was ground into a powder, the collected powder was then calcined at 350 ℃ at 3℃/min for 1 h in argon.

2.2 Characterization

The morphology of V_6O_{13} were observe by Hitachi S-4800 Field emission scanning electron microscopy (FESEM).The phase of V_6O_{13} were obtained by Panalytical X'Pert PRO MRD X-ray diffraction (XRD). X-ray photoelectron spectroscopy (XPS) analysis performed by Thermo Electron Corporation ESCALAB 250Xi. Energy dispersive spectrometer (EDS) by Oxford Instruments INCAIE 350 was used to analyze component of V_6O_{13}.

DOI: 10.1051/mateccconf/20178801006

3 Results and discussion

Through X-ray diffraction (XRD) the as-prepared samples were characterized firstly. Figure 1illustrated the XRD patterns of both pure V_6O_{13} and Al/Cr-doped(0.02, 0.06) V_6O_{13} powders calcined at 350℃.In the graph, no evidence of impurities for the three samples was detected from the XRD curves according to the standard card No. 71-2235[2]. For pure V_6O_{13}, there were main characteristic peaks at 2θ=15.122, 25.349, 26.842, 30.131, 33.487, 45.619, 49.496, 59.795 and 69.500, corresponding to (200), (110), (003), (-401), (310), (-601), (-603), (-711) and (025) planes, respectively, which can be indexed to JCPDS card No. 71-2235. For the other two Al/Cr-doped V_6O_{13}, all the diffraction peaks were also in accordance with the standard diffraction peaks of pure V_6O_{13}, which meant that did of Al/Cr doping did not change the basic structure of V_6O_{13}. Furthermore, Lattice constant may changed by dope because the ionic radius of Al^{3+} (0.535 Å) and Cr^{3+} (0.755Å) are differently the ionic radius of V^{4+} (0.58 Å) and V^{5+} (0.54 Å).

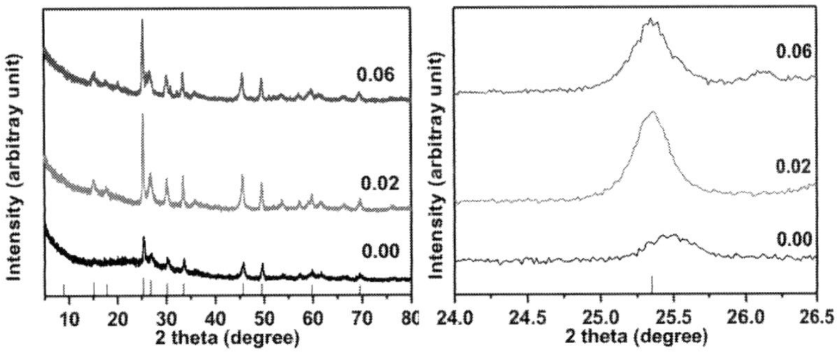

Figure 1. The XRD patterns of both pure V_6O_{13} and Al/Cr-doped(0.02、0.06) V_6O_{13} powders calcined at 350℃.

In order to ascertain actual content in V_6O_{13}, samples were tested by X-ray energy dispersive spectroscopy(EDS). Table 1shown the mass ratio of vanadium, oxide, aluminum, chromium with different quantities Al/Cr-doped V_6O_{13}. The amount of vanadium were decrease with Al/Cr doping V_6O_{13} and the amount of vanadium were continued decrease with the increase of doping. This was because the position of the aluminum and chromium replaced the vanadium in V_6O_{13}.

Figure 2 shown the SEM images of pure V_6O_{13} and Al/Cr-doped V_6O_{13} samples fabricated by hydrothermal method. In the figures, it can be easily seen that the structure units of pure V_6O_{13} and Al/Cr-doped were nanosheets. The doping amount of Al/Cr was demonstrated to have a effect to the samples final morphologies. Pure V_6O_{13} agglomeration serious. Al/Cr-doped are adhered together by stacking region in an regular arrangement with every unit. The thickness of pure V_6O_{13} nanosheets was about 500-1000nm larger than Al/Cr-doped V_6O_{13} and the thickness of Al/Cr-doped V_6O_{13} decrease with the increase of doping (0.02 about 400nm, 0.06 about 250nm). Besides, compared with the pure V_6O_{13}, Al/Cr-doped V_6O_{13} sample have more space this meant that it can accommodate much more lithium-ion(Li^+) when charge and discharge.

The cyclic voltammetry curves of sample 0.00, 0.02, 0.06 were shown in Figure 3a. The cyclic voltammetry curves of sample 0.00, 0.02, 0.06 were characterized at scan rate 0.1 mV s^{-1} and the scope of voltage was 1.5-3.5 V. For pure V_6O_{13} four oxidation peaks occurred at around 2.35, 2.68, 2.84 and 3.31 V (versus Li^+/Li), which represented the intercalation of Li into the non-equivalent sites in the structure of V_6O_{13}. While the reduction peaks located at around 2.04, 2.44 and 2.71 V (versus Li^+/Li) corresponds to the deintercalation of Li from the monoclinic V_6O_{13}. For Al/Cr-doped V_6O_{13}, the oxidation peaks of 0.02 centered at 2.43, 2.81, 2.93and 3.28 V while the reduction peaks of 0.02 centered at 1.97, 2.37 and 2.66 V, the oxidation peaks of 0.06 centered at 2.43, 2.81 and 3.35 V while the reduction peaks of 0.06 centered at 1.97, 2.40 and 2.68V, respectively. After doped with different amounts of Al/Cr, the oxidation peak position and restore its peak there were different change, indicating that the incorporation of Al/Cr changed vanadium oxide of vanadium ions in the Fermi level.

Table 2. The chemical composition of sample 0.00, 0.02, 0.06.

Sample	V (mass ratio)	O (mass ratio)	Al(mass ratio)	Cr(mass ratio)
0.00	61.66	38.34	0	0
0.02	47.93	50.32	1.23	0.53
0.06	42.00	55.26	1.15	1.64

Figure 2. The SEM images of pure V_6O_{13}(0.00) and Al/Cr-doped V_6O_{13}(0.02, 0.06).

Figure 3b sample 0.00, 0.02, 0.06 after three cycles of electrochemical impedance spectroscopy. After fitting, pure V_6O_{13} charge transfer resistance after 3 cycles was 151.6 Ω, and charge transfer resistance of the sample 0.02, 0.06 after 3 cycles are 383 Ω, 493 Ω, respectively. V_6O_{13} doped with Al/Cr were greater than the charge transfer resistance of pure V_6O_{13}, charge transfer resistance increases with the increase of the amount of doping, which may be V_6O_{13} spacing of layers increases and partially substituted lithium Al^{3+}/Cr^{3+} position, thus preventing insertion of lithium ions, so that impedance increases.

DOI: 10.1051/matecconf/20178801006

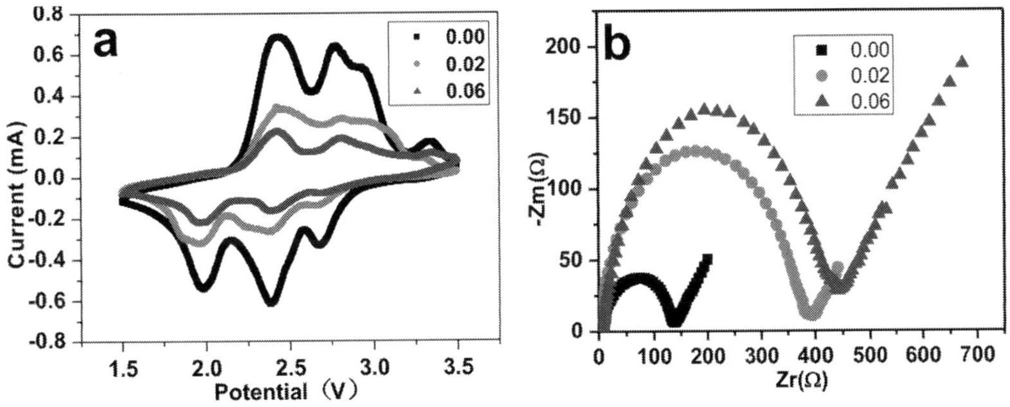

Figure 3a. The cyclic voltammetry curves of the third cycle for the pure V_6O_{13}(0.00) and Al/Cr-doped V_6O_{13}(0.02, 0.06). **Figure 3b.**Electrochemical impedance spectroscopy for samples 0.00, 0.02, 0.06 after the 3rd cycle.

In order to measurement the property of Al-doped V_6O_{13} anode material for lithium ion battery, cycle of button cells at room temperature. Figure 4 shown that the relationship between capacity and cycle number. The charge and discharge curves at the current density of 0.1 A g^{-1} in 1.5–4.0 V. All doping samples V_6O_{13} discharge performance were better than pure V_6O_{13}, initial discharge specific capacity sample 0.02 and 0.06 were 311mAh/g and 337mAh/g above pure V_6O_{13} sample was 241 mAh/g. The capacity retention of samples 0.00, 0.02, 0.06 after 100 cycles was 32.0%, 44.69%, 28.78%, respectively. The increase in capacity was attributed to introduction of Al/Cr, Al/Cr add to V_6O_{13} could increase electrical conductivity, nevertheless a great amount of Al/Cr could destroy V_6O_{13} crystal structure cause capacity declined sharply.

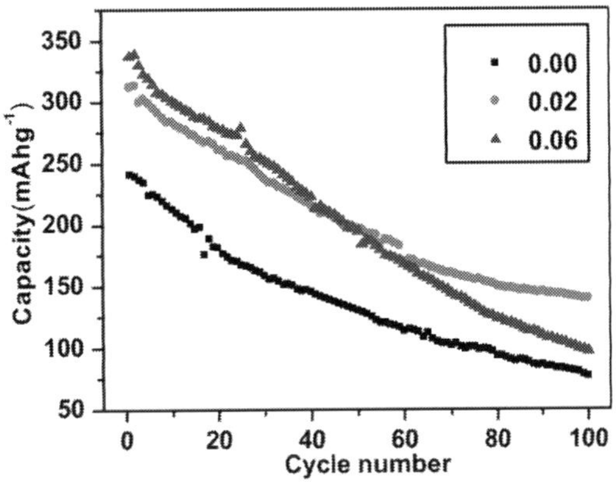

Figure 4. Cyclic performance of the pure V_6O_{13} (0.00) and Al/Cr-doped V_6O_{13} (0.02, 0.06) electrodes at 0.1 C.

4 Conclusion

Pure V_6O_{13} and Al/Cr-doped V_6O_{13} were synthesized via a completely aqueous solution based synthesis method. Doping proven to be an effective means to improve the cathode material charge and discharge specific capacity. Lattice constant may change by dope because the ionic radius of Al^{3+} and $Cr^{3+}$are differently the ionic radius of V^{4+} and V^{5+}. The doping amount of Al/Cr is demonstrated to have a effect to the samples final morphologies. the thickness of Al/Cr-doped V_6O_{13} decrease with the increase of doping. All doping samples V_6O_{13} discharge performance are better than pure V_6O_{13}, initial discharge specific capacity sample 0.02 and 0.06 are 311mAh/g and 337mAh/g above pure

V_6O_{13} sample is 241 mAh/g. The capacity retention of samples 0.00, 0.02, 0.06 after 100 cycles is 32.0%, 44.69%, 28.78%, respectively.

References

1. J. Haber, M. Witko, R. Tokarz. APPL CATAL A-GEN. **157**, 3-22(1997).
2. N. Peys, Y. Ling, D. Dewulf, S. Gielis, C. D. Dobbelaere, D. Cuypers, P. Adriaensens, S. V. Doorslaer, S. D. Gendt, A. Hardy and M. K. V. Bael. DALTON T. **42**, 959-968(2013).
3. Y. Y. Xia, T. Fujieda, K. Tatsumi, P. P. Prosini and T. Sakai. J.Power Sources **92**, 234-243 (2001).
4. J. Barker, R. Koksbang. Electrochim. Acta. **40**, 673-679(1995).
5. J. Barker, E. S. Saidi and M. Y. Saudi. Electrochim. Acta. **40**, 945-952(1995).
6. H. BjÖrk, S. Lidin, T. Gustafsson and J. O. Thomas. Acta Crystallogr, Sect. B: Struct. Sci. **57**, 759-765(2001).
7. J. HÖwing, T. Gustafsson and J. O. Thomas. Acta Crystallogr., Sect. B: Struct. Sci. **59**, 747-752(2003).
8. P. Soudan, J. P. Pereira-Ramos, J. Farcy, G. Gregoire and N. Baffier. Solid State Ionics. **135**, 291-295(2000).
9. C. Leger, S. Bach and J. P. Pereira-Ramos. J. Solid State Electrochem. **11**, 71-76(2007).

Deposition Time and Thermal Cycles of Fabricating Thin-wall Steel Parts by Double Electrode GMAW Based Additive Manufacturing

Dongqing Yang[1] and Guangjun Zhang[a,1]

[1]*State Key Laboratory of Advanced Welding and Joining, Harbin Institute of Technology, Harbin 150001, China;*

Abstract. The deposition time for fabricating the thin-wall part as well as the peak temperature of the substrate during the process was analyzed in the double electrode gas metal arc welding (DE-GMAW) based additive manufacturing (AM). The total deposition time and the interlayer idle time of the manufacturing process decreased with the increasing of the bypass current under the same interlayer temperature and the same deposition rate. The thermal cycling curves illustrated that the peak temperature of the substrate was lower in the DE-GMAW base AM under the same conditions. When depositing the thin-wall parts, the DE-GMAW based AM can reduce the heat input to the substrate and improve the fabrication efficiency, compared with the GMAW based AM.

1 Introduction

Additive manufacturing (AM) is an advanced fabrication technique that can make parts through a layer by layer deposition based on the virtual solid models. Due to its considerable reductions in production costs and savings of materials, AM has become a hotspot of research in the worldwide [1]. Especially, with the increasing requirement for the industrial technologies, fabricating metallic functional products by additive manufacturing has attracted the attention of many researchers over the recent years [2]. According to the heat source utilized in the deposition process, metallic components AM technologies can be divided into following three categories mainly: laser, electron beam and welding arc. The equipment expenditure and maintenance costs were high in laser and electron beam based AM. In contrast, the welding arc based AM has the advantages of the simpler equipments and low production cost [3]. Moreover, the metal wire was employed as the deposited raw materials in mostly welding arc based AM, thus it was a cleaner technique with fewer contamination issues, compared with the process using powder [4]. Based on the different welding processes using in AM, it can be classified into following categories: gas metal arc welding (GMAW), gas tungsten arc welding (GTAW), and plasma arc welding (PAW). In GTAW and PAW based AM, the wire was fed into the molten pool individually and the arc was ignited between the tungsten and the base metal. But in GMAW based AM, the welding arc was directly struck between the wire and the substrate. In this condition, the electric arc will melt the wire quickly and it will lead to high deposition rate in GMAW base AM. Thus, GMAW based AM is more suitable for fabricating the medium and large scale structure metal parts [5].

[a] Corresponding author : zhanggj@hit.edu.cn

A number of investigations about the GMAW-based AM have been explored recently [6-10]. However, during the traditional researches on GMAW-based AM, the heat accumulation in the deposition is a common problem that restricts the development and application of GMAW based AM technique. The usual measures to solve this problem are decreasing the wire melting current, increasing travel speed or delaying the time between two adjacent layered depositions. Those will lead to lower deposition rate and weaken the advantages of this technique. Taking a page from the welding field, the double electrode gas metal arc welding (DE-GMAW) technique proposed by Zhang et al. and improved by Li et al. was characterized by reducing the heat input into the substrate through the bypass arc under the same deposition rate [11-12]. It was introduced into the weld-based additive manufacturing in our previous research [13]. The forming characteristics in multi-layer single-bead additive manufacturing by DE-GMAW were investigated and the effect of the bypass current on the deposition quality was illustrated in that work. Based on it, this paper aims at analyzing the thermal behavior of the DE-GMAW based AM process by the measurement of the thermal cycles and investigating the effect of the bypass arc on the deposition time and the thermal cycles.

2 Experimental procedure

The DE-GMAW based additive manufacturing system is shown in Fig. 1. DE-GMAW was established by introducing a GTAW torch to form a second current loop for the wire melting current. In that way, the base metal current was decoupled from the melting current and heat input to the base metal was lower. In this work, the Panasonic YD-500FR welding machine with constant voltage mode was the GMAW power supply and the Rilon WS-400 with constant current mode was the bypass arc power supply. The computer can achieve the following functions: controlling the movement of the work flat, on/off control of GMAW and GTAW, adjustment of relative process parameters, and collecting and recording the data of the thermal cycles during the deposition. In this work, both welding torches were always static as shown in Fig. 1 and the thin-wall part was fabricated with the descending of the work flat. The H08Mn2Si steel wire with 1.2mm diameter was served as consumable material deposited on the Q235 low carbon steel plates with dimension of 250mm×80mm×9.5mm. The diameter of the tungsten electrode used in the GTAW torch was 3.2mm. The shielding gas for GMAW torch was Ar (95 %) and CO_2 (5 %) gas mixture with a constant flow rate of 15 L/min and the pure Ar gas was supplied for the GTAW torch with 8L/min flow rate. The relative position of two welding torches was just same with Ref. [13].

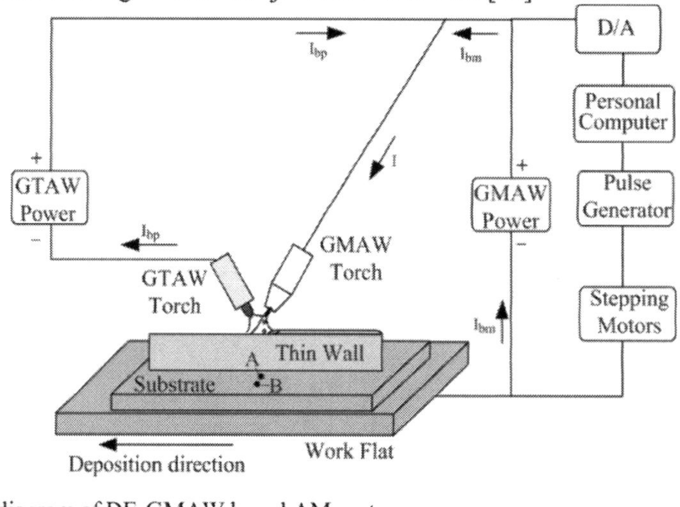

Figure 1. Schematic diagram of DE-GMAW based AM system.

For this research, the deposition time and thermal behavior of the DE-GMAW based additive manufacturing are key points. Based on the previous researches, we have conducted some groups of experiments with depositing thin-wall parts. As illustrated in Tab.1, the experimental parameters were listed in detail. In the experiment 1-4, the every layer has been deposited 180mm length and the interpass temperature of the deposited parts was controlled at 70℃ measured in the middle of the top surface by an infrared thermometer. The interval time and the total deposition time were recorded. In the experiment 5-6, the thermal cycling data of the deposition process has been collected, the deposited length of every layer was 150mm, and the interpass temperature was 60℃. The thermal cycling curves on substrate were measured by some thermocouples (K type) distributing at points A and B on the top surface of substrate as shown in Fig.1. The distance from Point A to the center line of the parts was 15mm and the Point B was 30mm away from the center.

Table 1. Experimental condition of DE-GMAW based AM.

No.	Wire feed speed (m/min)	Bypass current (A)	Arc voltage (v)	Deposition velocity (mm/s)	Total number of layers
1	4.21	0	23	5	20
2	4.21	40	23	5	20
3	4.21	80	23	5	20
4	4.21	120	23	5	20
5	4.21	0	23	5	15
6	4.21	80	23	5	15

3 Results and discussion

3.1 Deposition time

In the Exp 1-4, except for the first layer, the other layers were deposited after the temperature of the previous layer metal dropping to 70℃. Thus, there was an interval between the adjacent layers, namely interlayer idle time. The interlayer idle time of every layer deposition for cooling the parts with different bypass current was shown in Fig. 2. It is obvious that the idle time between the adjacent layers decreased with increase of the bypass current in general. It was due to the reduction in the heat input of the deposited part with the bypass arc. The higher bypass current caused more reduction of the heat input in DE-GMAW based AM. In each experiment, the time interval between the contiguous depositions was shortest between the first layer and second layer, then it was extended as the increasing of the deposition layer number, and finally it reached a certain value with small fluctuation. This was because the first layer deposition was closed to the cool substrate with good heat diffusion condition and the temperature of the metal layer dropped fast. With the increasing of deposition height, the heat diffusion condition was poor and the time interval for cooling the part was stretched and maintained stable eventually.

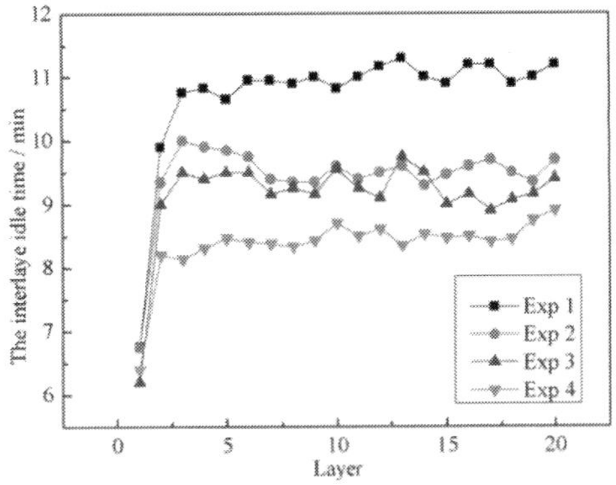

Figure 2. The interval time between the adjacent layers in the Exp 1-4.

 DOI: 10.1051/matecconf/20178801007

In Tab.2, the total deposition time, total interlayer idle time and mean interlayer idle time were listed. The total deposition time included the interlayer idle time and the actual deposition time. In Exp 1-4, the actual deposition time was always 12min according to the calculation. Therefore, the total deposition time depended on the interlayer idle time of every layer deposition. As illustrated in Tab.2, the total interlayer idle time decreased with the increase of the bypass current. The maximum reduction of the total deposition can reach 21% with the bypass current of 120A. The total volume of deposition metal was constant in theory due to the same wire feed speed, deposition velocity and deposition length in the Exp 1-4. The more the total deposition time, the lower the production efficiency. That is to say, the manufacturing efficiency during the DE-GMAW based AM was higher than the one of the GMAW based AM with the same conditions, taking no account of the deposition quality. In the Fig.3, the deposition time of different layer number was always shortened with the growth of the bypass current. The larger the layer number, the more the shortened time of the deposition in DE-GMAW based AM, compared with the GMAW based AM in the same deposition rate.

Table 2. The total deposition time in experiment 1-4.

Experiment	Total deposition time (min)	Total interlayer idle time (min)	Mean interlayer idle time (min)
1	226.39	214.39	10.72
2	200.43	188.43	9.42
3	194.52	182.52	9.13
4	179.22	167.22	8.36

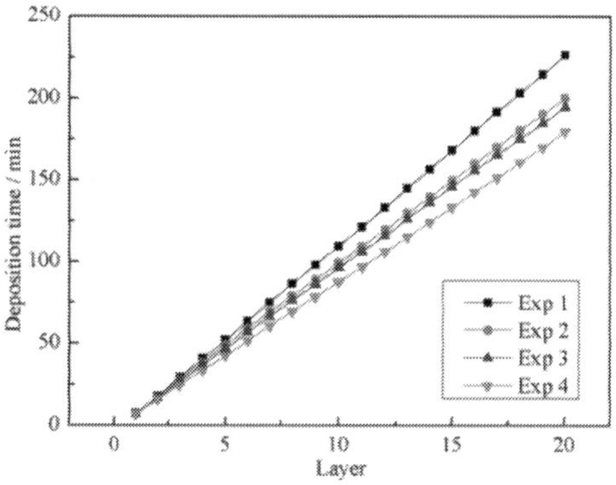

Figure 3. The deposition time with different layer numbers in Exp 1-4.

3.2 Thermal cycling

The thermal cycles on the substrate with different parameters have been acquired in the Exp 5-6. The typical thermal cycles were shown in Fig. 4 and the characteristics of the curves were obvious. Here just listed the thermal cycling curves of the point A and point B on the substrate. The welding arc as the heat source was applied on the parts 15 times in every experiment. Accordingly, the points of the substrate went through thermal cycling with 15 peaks in this condition. With the increase of layer number, the heat source was farther from the points on the substrate. As a result, the peak temperature of the points declined gradually. Meanwhile, the interlayer temperature of each layer was controlled at 60℃ and the temperature of the point A on the substrate was slightly lower about 55℃. The point B was farther away from the center of the part than the point A and the peak temperature was lower. The cooling curves of the thermal cycling were mainly synchronized in both points.

DOI: 10.1051/matecconf/20178801007

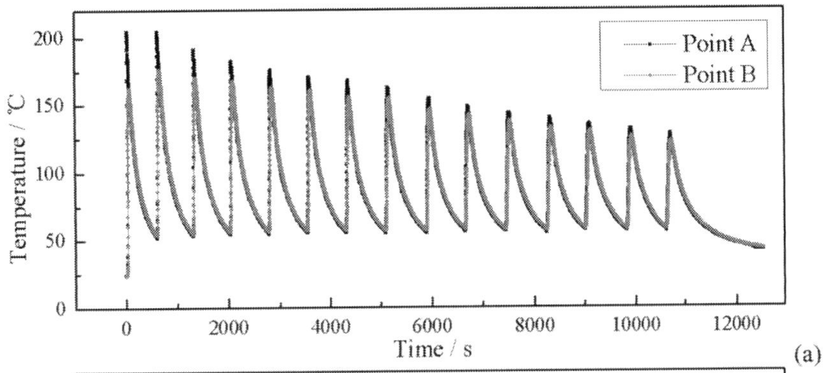

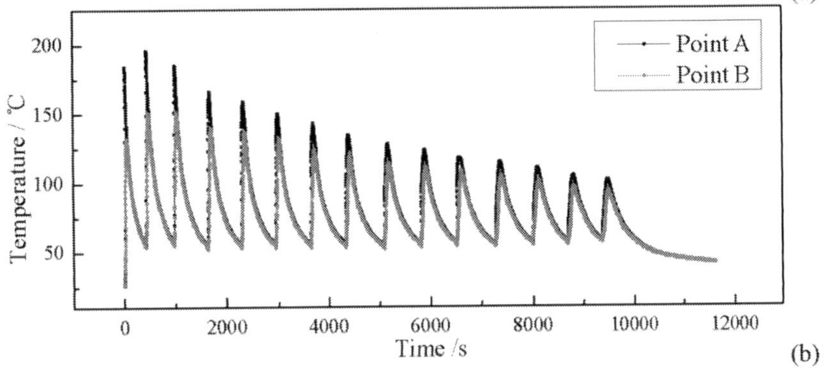

Figure 4. Thermal cycles of Point A and B in Experiment 5-6. (a) Exp 5. (b) Exp 6.

The effect of the bypass arc on the thermal cycles of the points on the substrate was distinct and can be seen in the Fig. 5. Thermal cycling curves of the points on the substrate were similar, except for the peak temperature. Thus, only the result of the point A was presented here. The peak temperature in Exp 5 of each layer deposition was higher than that with bypass current in Exp 6. That means the heat input during the GMAW based AM is higher than that during the DE-GMAW based AM with suitable bypass current in the same deposition rate. The result of the total deposition time and interval idle time in different experiments was in accord with the content of Section 3.1.

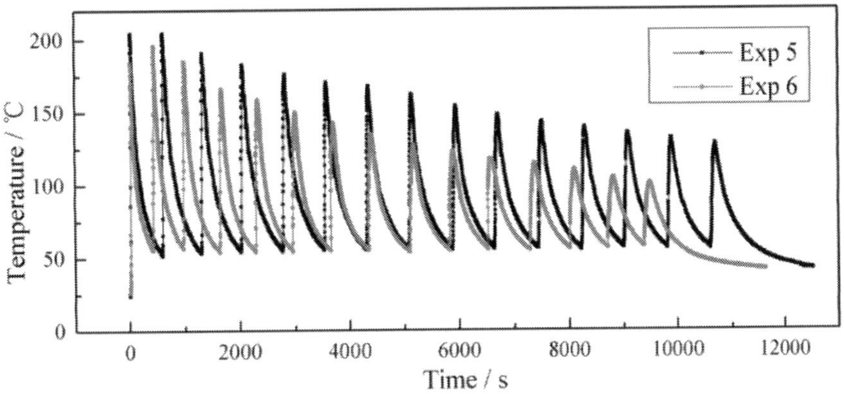

Figure 5. Comparison of thermal cycles of Point A in Exp 5 and Exp 6.

4 Conclusions

The total deposition time of fabricating the thin-wall part and the interval time decreased with the increasing of the bypass current under the same deposition rate during the DE-GMAW based AM. The peak temperature of the substrate during the deposition was lower in the DE-GMAW based AM compared with the one in GMAW based AM. When depositing the thin-wall parts, the DE-GMAW based AM can reduce the heat input to the substrate and improve the fabrication efficiency, compared with the GMAW based AM.

References

1. S.H. Huang, P. Liu, A. Mokasdar, L. Hou, Int. J. Adv. Manuf. Technol. **67,** 1191-1203 (2013).
2. G. Tapia, A. Elwany, ASME J. Manuf. Sci. Eng. **136**, 1-10 (2014).
3. B. Baufeld, E. Brandl, O. Biest, J. Mater. Process. Technol. **211**, 1146-1158 (2011).
4. F. Martina, J. Mehnenb, S.W. Williams, P. Colegrove, F. Wanga, J. Mater. Process. Technol. **212**, 1377-1386 (2012).
5. M.P. Mughal, H. Fawad, R.A. Mufti, Proc. Inst. Mech. Eng. Part C-J. Eng. Mech. Eng. Sci. **220**, 875-885 (2006).
6. H. H. Zhao, G.J. Zhang, Z.Q. Yin, L. Wu, J. Mater. Process. Technol. **211**, 488-495 (2011).
7. P.A. Colegrove, H.E. Coules, J. Fairman, F. Martina, T. Kashoob, H. Mamash, L.D Cozzolino, J. Mater. Process. Technol. **213**, 1782-1791 (2013).
8. P. Kazanas, P. Deherkar, P. Almeida, H. Lockett, S. Williams, Proc. Inst. Mech. Eng. Part C-J. Eng. Manuf. **226**, 1042-1051 (2012).
9. X.W. Bai, H.O. Zhang, G.L. Wang, Procedia CIRP. **6**, 515-520 (2013).
10. J. Xiong, G.J. Zhang, Z.L. Qiu, Y.Z. Li, J. Clean. Prod. **41**, 82-88 (2013).
11. Y.M. Zhang, M. Jiang, W. Lu, Weld. J. **83**, 39-41 (2004).
12. K.H. Li, , J.S. Chen, Y.M. Zhang, Weld. J. **86**, 231s-237s (2007).
13. D.Q. Yang, C.J. He, G.J. Zhang, J. Mater. Process. Technol. **227**, 153-160 (2016).

MATEC Web of Conferences **88**, 01008 (2017) DOI: 10.1051/matecconf/20178801008
CBNCM 2016

A Fast and Efficient Dehydration Process for Waste Drilling Slurry

Guo Zheng[1,a], Cailin Wang[1] and Yu Sun[1]

[1]*School of Environment and Chemical Engineering, Tianjin Polytechnic University, Tianjin 300387, China*

Abstract. In this article, slurry system was converted to colloid from fluid with the colloidization of high polymer coagulants with high viscosity. The solid-liquid separation of the waste slurry was realized by the process of chemical colloidal gel breaking, coagulation function, acidification gelout. In addition, the surface morphology of slurry cake was investigated by using Field emission scanning electron microscope (FE-SEM). The results indicate that mud separation effect is decides on the type of flocculants, gel breaker. The solid content of mud cake increases from 40.5% to 77.5% when A-PA and H20 are employed as the flocculants, gelout, with the dosage of zero point four grams and zero point five grams.

1 Introduction

Drilling sludge, as a key material in the oil and gas well drilling operation, is used to flush wellhead[1-2]. However, according to the survey, the annual production of waste drilling sludge is increasing trend year by year in China[3]. Without processing direct emissions, it cause serious environmental problems[4-6]. Because the waste drilling mud contains a large number of organic and inorganic pollutants, without processing direct emissions, it not only takes up a lot of land area, soil compaction and arbitrary term deposits, some harmful substance by rain flushing would leak into the surrounding soil, vegetation, livestock and so on. At present, the chemical solid-liquid separation method is more, through the mud to join the breaker to remove the mud, through mechanical means to achieve solid - liquid separation[7]. This article through the electrical repulsion to further strengthen the colloidal properties of the mud, the fluid completely transformed into jelly, using colloidal gel breaking, poly sink properties to achieve a slurry of solid - liquid separation.

2 Experimental materials and methods

2.1 Materials

Polymer coagulating agent A-PA (anionic), C-PA (cationic), N-PA (non-ionic), industrial pure, Tianjin university of technology textile auxiliaries co, LTD.; H20 , industrial pure, Tianjin university of technology textile auxiliaries co, LTD.; sulphuric acid, nitric acid, hydrochloric acid and boric acid, hydrogen peroxide, Tianjin North Asclepius chemical reagent factory; calcium chloride, aluminum

[a] Corresponding author: zhengguo0703@126.com.org

sulphate, poly-aluminium chloride, calcium hydroxide, ferric chloride, analysis of pure. Tianjin Kermel Chemical Reagent Co. LTD.

2.2 Equipment

Analytical balance, FA2004N, Shanghai Jing marine instrument Co., Ltd.; PHS-25, Shanghai Precision Scientific Instrument Co., Ltd.; circulating water vacuum pump, SHZ-D (III) type, Gongyi City, granting China Instrument Co., Ltd.; precision timing electromotive stirrer, jj-1, Jintan City Liangyou instrument is limited; field emission scanning electron microscopy, Quanta 200, Fei company, chemical oxygen demand (COD) rapid determination instrument, 5B-3A (V7) type and even China Lanzhou Environmental Protection Technology Co., Ltd.; inductively coupled plasma atomic emission spectrometry (ICP-AES), Varian 715-ES, spectro analysis instrument company.

2.3 Mud pretreatment

Due to the long time, some performance of the waste drilling sludge is influenced by the gravity. For example, the rheological property becomes weak, the colloid ratio decreases, the precipitation rate increases, and the continuous network structure of the lower sludge is much stronger than that of the continuous net structure of the upper sludge. In order to ensure the consistency of the overall performance of the sludge, mixing must be carried out before use.

2.4 Slurry solid-liquid separation process

It takes one hundred take 100 gram of pretreatment of waste drilling mud, adding organic coagulant, mixing, stirring speed was 200 revolutions per minute, until the mud together to form a group and no longer adhere to the wall of the beaker and the treatment of colloidal mud jelly, no liquidity. Then it does to break the glue, which needs to add the gel breaker, shear type stirring, stirring speed of 300-350 revolutions per minute, until the existence of no large block of plastic. Coagulant aid: adding coagulant, shear type stirring, stirring speed of R 150-200 revolutions per minute, to small particles and large particles of suspended particles. The vacuum filtration time is 20-30min, and the vacuum pressure maintained is at 0.08-0.09MPa. All of the above process temperature is 25 degrees centigrade, the relative humidity is 50%.

2.5 Preparation of mud cake leaching solution and determination of COD and heavy metal ions

Reference GB5086.2-1997solid waste leaching toxicity level oscillation method and execute GB5085.3-2007hazardous waste identification standard Leaching toxicity identification.
Under natural conditions, the solid state of the waste drilling mud after solid liquid separation is achieved in a static state. Therefore, take 100 gram (dry weight), placed in the 2 litre polyethylene wide mouth bottle (with a sealing plug), adding 1000 milliliter de ionized water, static soak 24 hours, take the supernatant for analysis COD. Finally, the supernatant was filtered through micro porous filter of 0.45 micron, and the filtrate was used for the analysis of heavy metal ions.

2.6 Field emission microscope

Take waste drilling mud sample field emission microscope to observe the morphology of the particles. After solid-liquid separation, it takes a small amount of sample to field emission microscope to observe the morphology of the particles.

DOI: 10.1051/matecconf/20178801008

2.7 Technological process

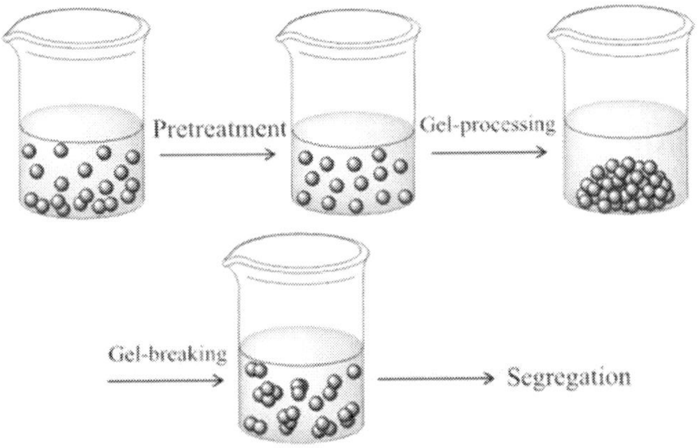

Figure 1. Process flow diagram of waste drilling mud on solid-liquid separation

3 Results and discussion

3.1 Influencing factors of solid-liquid separation of waste drilling mud

3.1.1 Effect of type of coagulant on the effect of solid liquid separation

Respectively we choose three different types of coagulants. They are A-PA, C-PA and N-PA. The separation effect is shown in figure 2.The type of flocculants is a decisive factor for mud solid-liquid separation process; with the increasing of dosage of C-PA, solid content of mud cake is less, which is a negative growth trend. But the using of A-PA significantly enhances the waste mud solid-liquid separation effect. Therefore the changing of N-PA dosing quantity does not affect the basic abandoned drilling mud solid-liquid separation effect. The main reason is that the charge neutralization makes mud dilute, losing colloidal properties. In alkaline condition, A-PA occurs hydrolysis. Some of its amide groups form negative charges, which repel solid phase particles of mud .Because the surface of solid phased particles in the mud is contained negative charge. Repulsive force between the molecules stretch makes molecular structure morphology of polymer flocculants stretch and increases Zeta potential increase of clay particles. The whole system gets into a more stable colloidal system on the apparent. It helps to take advantage of the settlement of colloid to realize solid-liquid separation performance. Therefore, in this solid-liquid separation process of slurry, we choose A-PA flocculants.

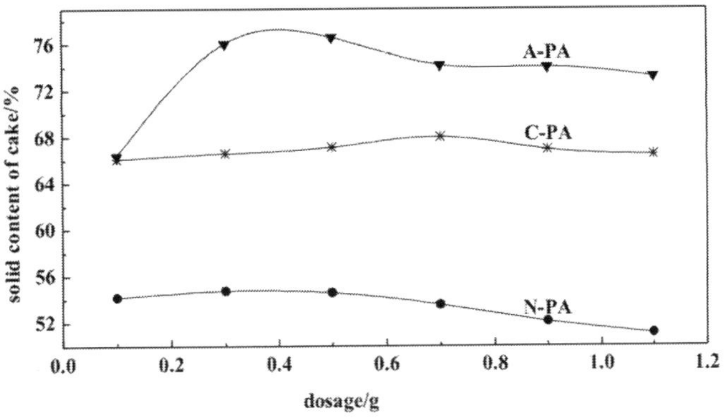

Figure 2. Effect of coagulant type on the separation.

3.1.2 Effect of A-PA dosage on the effect of solid liquid separation

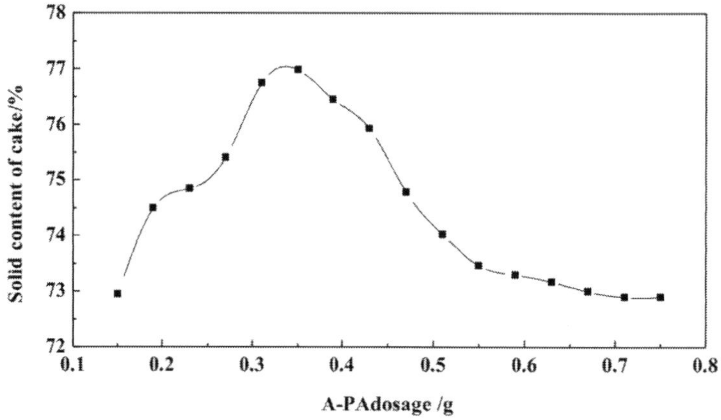

Figure 3. Effect of A-PA dosage on the separation T.

Figure 3 shows the effect of A-PA dosage on the solid liquid separation. With the increasing of A-PA dosage, the solid content of the cake increases first and then decreases. This is mainly because some free water reform to bound water in the mud and cannot realize effective separation. And with the increasing of A-PA, the surface of mud solid particles will adsorb amounts of high molecular material, and then form a layer of protective film in the solid particles, Therefore, in solid-liquid separation process of slurry, the optimal dosage of A-PA is 0.375g and the solid content of the mud cake reached the maximum value of 80.5%.

3.1.3 Effect of gel breaking agent on the effect of solid liquid separation

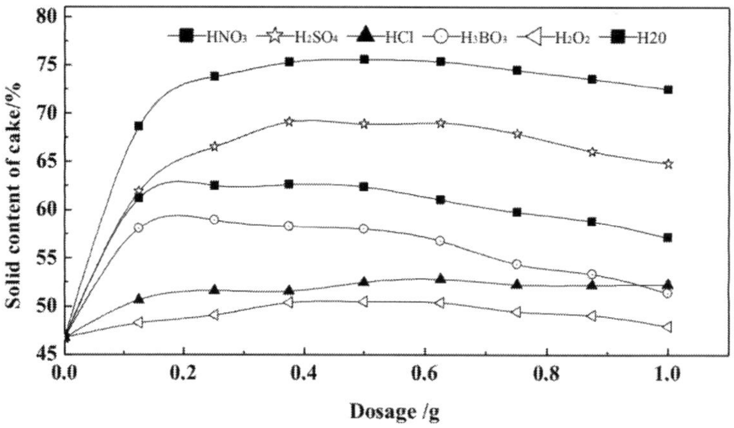

Figure 4. Effect of acidifier dosage on the separation

The effect of adding amount of gel breaker on the solid-liquid separation is shown in Figure 4. The addition of gel breakers is beneficial to the solid-liquid separation of slurry, but the separation effect is different. Among them, the effect of H20 is best. This main reason is that drilling mud is strong alkaline. And in the alkaline environment, oil and mud treatment agent can combine together and form high stability. By adding acid gel breaker, it can happen neutralization reaction and form complex salt, which changes the balance of the original system and destructs electric double layer. However, nitric acid, sulfuric acid and hydrochloric acid belong to strong acid, which are corrosive, they not only can

DOI: 10.1051/matecconf/20178801008

neutralize alkaline and destroy part of waste drilling mud lattice structure of clay particles, reducing the solid-liquid separation effect. Boric acid and hydrogen peroxide are weak acid, which can't completely break the glue. The acid strength of H20 is in middle position. And the best dosage of H20 is 0.5g.

3.2 Micro morphology of waste drilling mud

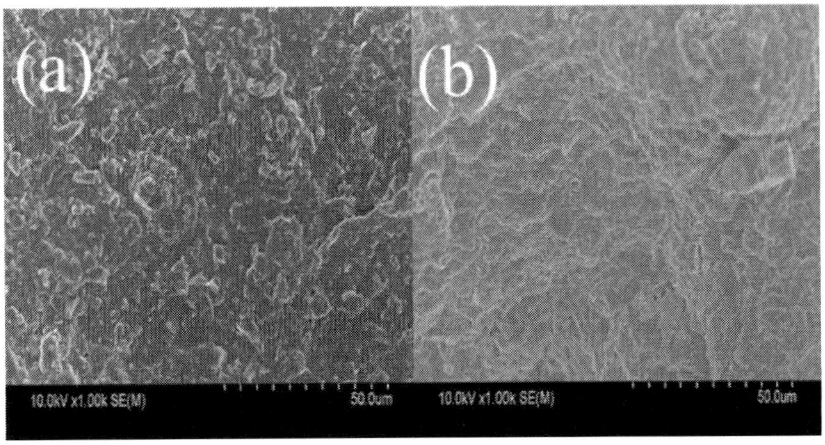

(a) mud cake before the solid-liquid separation (b) mud cake after the solid-liquid separation

Figure 5. SEM micrographs of mud cake

The micro morphology of the waste drilling slurry before and after solid-liquid separation is shown in Figure 5(a, b). Figure 5(a) shows that waste drilling mud contains different particle size of solid particles and oil drops. There are a lot of free water, solid phase particles scattered irregular, no obvious means of gathered phenomenon. The figure 5(b) shows that the mud almost contains litter free water in the mud cake after solid-liquid separation. And solid-phase material arranges closely and gathers big particles. Because the waste drilling mud treatment of moisture content is about 55.5%, all kinds of solid particles, oil drops and water are unable to form a uniform system, a mixed system of water in oil and oil in water.

3.3 Comparison of slurry indexes before and after solid-liquid separation

Table 1. The basic properties of waste drilling mud

Measure	Solid content/%	Volume /mL	PH	COD / $mg \cdot L^{-1}$	Total chromium /$mg \cdot L^{-1}$	Total lead /$mg \cdot L^{-1}$
Before	40.5	79.2	10.89	22564	0.0271	0.0969
After	77.5	31.5	6.72	88.7	0.0085	-

Table 1 shows that pH, COD and heavy metal content are significantly reduced after solid-liquid separation and meet the national emission requirements. Mud cake volume is more greatly reduced than the mud volume before separation, which can come true the effective realization of solid waste reduction of emissions and saving cost. Because during the slurry solid-liquid separation process, effective the harmful substances in the original slurry enrichment in the separation of the liquid phase, and after the separation of the liquid phase can be further processing and recycling through the film processing techniques.

4 Conclusions

(i) Mud separation effect by coalescing agent, acidulant and coagulant types and dosage, when the organic gelling agent for A-PA and gel breaker for H20, and dosage respectively for zero point four grams and zero point five grams, drilling mud solid-liquid separation effect best, mud cake of solid content by 40.5% increased to 77.5%.

(ii) Through the indoor experiment, pH, COD, heavy metal content reached the national emission requirements, volume decreased by approximately 60%, maximum realized solid waste reduction of emissions, with good economic and social benefits.

References

1. M. J. Yang, W. L. Liang, Q M Jing, et al. JMP, 23: 109-112,(2003)
2. G. Li, M Zhu, J. L. Qian. IWT, 17: 13-14,17,45,(1997)
3. J. Liu, J. Zhang. EE, 27: 91-93,(2009)
4. J. Q. Chen. SP, 399-402,(2005)
5. G. J. Hu, J. B. Li, G.M. Zeng. JHM, 261C: 470-490,(2013)
6. A. H. Long, Y. X. Sun. WCEE, 15: 165-168,(2003)
7. L. M. Guo,. J. Yuk, X. Chen. EPC, 37:58-61, (2015)

Comparison Between Different Ways in Making Silicon Dioxide Layer on Silicon Wafers

Chenguang Sun[1,a], Yanjun Wang[2], Qiang Xu[2], Xuenan Zhang[2] and Zhenfu Liu[1]

[1]*Tianjin Zhonghuan Advanced Material & Technology Co., Ltd, China, Tianjin City, 300384*
[2]*Tianjin HuanOu Semiconductor Material and Technology Co., Ltd. China, Tianjin City, 300384*

Abstract. The most important raw material for power device is epitaxial wafers, which is made from heavily doped polishing silicon wafers. In order to prevent dopant doped into silicon while pulling ingots spread into the epitaxial atmosphere and finally into the Epitaxial layer, while polishing wafers are produced, silicon dioxide layer will be added on the back surface of silicon wafers. There are two ways mainly used to make this silicon dioxide layer, one is named High Temperature Oxide (HTO), and the other is named Atmosphere Pressure Chemical Vapor Deposition (APCVD). The process parameter such as temperature, time and pressure are quite different from each other, also the character of the silicon dioxide layer and the effect that silicon dioxide layer have on the wafers are different from each other. In this article we will compare the difference between the two silicon dioxide layers and the ways in making silicon dioxide layers.

1 INTRODUCTION

Power Semiconductor is also known as Power Electronic Device, including Discrete Device and Integrated Circuit, mainly used in transforming and controlling current, voltage, frequency and so on in order to achieve the function as rectification, inversion, chopping, amplifier etc. Because these kinds of devices can endure high voltage large current, they play more and more important roles in nowadays society[1]. Together with the improvement of Power Semiconductor the requirement of the main raw material for power semiconductor is becoming more and more strict, also the requirements of polishing wafers is becoming more and more strict.

Epitaxial process is a high temperature process. During this process substrate character will have a lot effect on epitaxial layer, such as defects and WARP data. So there are special requirements while polishing wafers are produced. Also during epitaxial process, the dopant will go into the atmosphere due to the high concentration of dopant in the wafer, from the part which is exposed to the atmosphere. This behavior will result in the contamination to the epitaxial layer, the unwanted dopant mixing into the epitaxial layer will lead to resistance deviation, and finally the Radium Resistance Variation (RRV) will exceed target value. This phenomenon is named Self-Doping[2-4]. In order to avoid this Self-Doping effect, while polishing wafer is produced, silicon dioxide is made on the back surface of wafers. This silicon dioxide will stop dopant from diffusing into the atmosphere to avoid Self-Doping

[a] Corresponding author : sunchenguang@semicon-ho.com

phenomenon, but at the same time while silicon dioxide is made, thermal process will also work on defects gathering and geometry varying[5].

There are two main ways to make this silicon dioxide layer, HTO and APCVD[6], we will discuss the advantage and disadvantage if we choose which way to make the silicon dioxide layer, and compare the effect each way we take to make the silicon dioxide layer to the wafers.

2 Theories about making silicon dioxide layer

HTO Theories: The equipment we use to make silicon dioxide layer by HTO method is High Temperature Dispersing furnace. Wafers insert vertical on the carrier in the groove, there is small space between wafers, and the carrier will be pulled into the tube. Inside the tube the temperature is already higher than 900℃, the wafers will be kept in this tube for more than 8 hours, during this period, oxygen-oxygen and hydrogen-oxygen will be connected to the tube one after another. Oxygen in the tube will result in oxidation on both surfaces of the wafers. While we have oxygen and hydrogen in the tube at the same time we will have water, water will help to increase the rate of the oxidation. Different thickness of the layer is controlled by adjusting oxidation time, temperature and other parameters. After the oxidation process, wafers will be taken out from the tube then remove the silicon dioxide on the front surface of the wafers, because we can't polish the wafers with silicon dioxide on the surface.

APCVD Theories: The equipment we use to make silicon dioxide layer by APCVD method is APCVD furnace. Wafers will be laid on the surface of carrier horizontally, the backside of wafer contact the surface of carrier. Then the carrier will wander into and out the tube, during the wandering silane and oxide will be flow in, under the temperature about 500℃ silane will decompose and oxidized by oxygen into silicon dioxide. This process will take only 5 minutes and because one side of the wafers is protected by the carrier, there is only side will have silicon dioxide on the surface, so after APCVD, wafers are ready to be polished.

3 Experiments and Results

3.1 Comparison of thickness, thickness homogeneity and compactness of silicon dioxide layer between HTO and APCVD

We set our target of the thickness of silicon dioxide layer at 5100Å, after each process we measured 5 wafers from each process, each wafers measure 5 points as is shown in the Figure1. The results of thickness measurements are included in Table1.

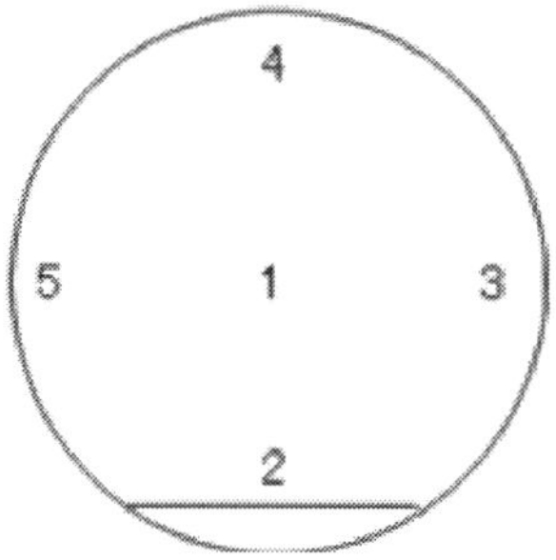

Figure1. Schematic of points chose to measure silicon dioxide layer thickness

From the table we find out that, HTO process is better than APCVD in homogeneity inside wafers and between wafers. The calculate formula is (max-min)/(2*average), from the result we can see homogeneity inside wafers from HTO process is always under 2%, the homogeneity between wafers in lower than 0.5%. The result shows by using HTO process we can get better homogeneity than APCVD. It is not strange to get this result, because HTO process is more stable than APCVD, a tube

 DOI: 10.1051/matecconf/20178801009

filled with oxygen, 8 hours high temperature process, also the silicon dioxide layer comes from the wafer itself, the process conditions for all the wafers are quite the same, there conditions will help to get better homogeneity in making silicon dioxide layer. But in the APCVD process, silicon dioxide layer is deposited on the surface of wafers, within 5 minutes, the wafers in wandering in the tube, also the airflow of silane and oxygen may change from time to time, these conditions will all result in worse homogeneity in silicon dioxide layer.

Table 1. Thickness and thickness homogeneity comparison between APCVD and HTO

	No.	1	2	3	4	5	AVE	MAX	MIN	VAR
APCVD	1	5103	5211	5056	5024	5032	5085.2	5211	5024	1.84%
	2	5167	5272	5209	5057	5110	5163.0	5272	5057	2.08%
	3	5116	5251	5064	5117	5060	5121.6	5251	5060	1.86%
	4	5178	5277	5214	5213	5112	5198.8	5277	5112	1.59%
	5	5024	5118	4830	5050	5035	5011.4	5118	4830	2.87%
	Variation between wafers:									1.83%
HTO	1	5095	5142	5103	5016	5110	5093.2	5142	5016	1.24%
	2	5030	5126	5069	4983	5110	5063.6	5126	4983	1.41%
	3	5073	5140	5090	5011	5115	5085.8	5140	5011	1.27%
	4	5069	5155	5110	5020	5120	5094.8	5155	5020	1.32%
	5	5040	5165	5090	5021	5110	5085.2	5165	5021	1.42%
	Variation between wafers:									0.31%

Because silicon dioxide is only dissolved in hydrofluoric acid (HF), so we can use diluted HF to etch the silicon dioxide, if the compactness of silicon dioxide is different, with the same thickness of silicon dioxide, the dissolving time will be different. We can use eye inspection to check if there is silicon dioxide remained on the surface of wafers or not, if silicon dioxide remains, the surface will be hydrophilic, else will be hydrophobic. In order to control the dissolve rate, we diluted 49% HF with water (H_2O_2) at 1:14, and then inserted wafers from APCVD and HTO into the cassette together, then dropped the cassette into the solution. After etching some time, pulled the cassette up and checked the surface to see if it is hydrophilic or hydrophobic, etching time and surface status are recorded in Table 2. From the table we can see that, silicon dioxide from APCVD is much easier to remove, within 2 minutes the silicon dioxide is partly removed, and within 3 minutes, the silicon dioxide is completely removed. At this minute, silicon dioxide from HTO is still covered the whole surface of wafer. Continued on the experiments, we can see that silicon dioxide from HTO is partly removed at 8 minutes, and totally removed within 10 minutes.

Table 2. Etching time and surface status of APCVD and HTO wafers

Etching time	HTO	APCVD	NOTES
1min	Hydrophilic	Hydrophilic	
1.5min	Hydrophilic	Hydrophilic	
2min	Hydrophilic	Partly hydrophobic	Silicon dioxide partly removed
3min	Hydrophilic	Hydrophobic	Silicon dioxide totally removed
4min	Hydrophilic	Hydrophobic	
6min	Hydrophilic	Hydrophobic	
8min	Partly hydrophobic	Hydrophobic	Silicon dioxide partly removed
10min	Hydrophobic	Hydrophobic	Silicon dioxide totally removed

DOI: 10.1051/matecconf/20178801009

Repeated the above experiment, we got the same result. It is very normal to get this result, because with HTO process, silicon atoms near the wafer surface will be oxidized into silicon dioxide, this means oxygen atoms go into the gap between silicon atoms, this will make the density of silicon dioxide layer very high. For APCVD process, silicon dioxide is formed by silane and oxygen, and then deposited on the surface of wafers, the density is much lower than HTO.

In this part, we tried different dopant such as Arsenic, Antimony, Boron and Phosphorous, different orientation <111> and <100>, we always have the same result, it means that thickness homogeneity and compactness of silicon dioxide layer will only decided by the thermal process, wafer character will not impact on these items.

From section 2.1 we will know that HTO do a better job than APCVD, we will get better thickness homogeneity and compactness of silicon dioxide layer by using HTO.

3.2 Comparison of geometry parameter and defect gathering between HTO and APCVD

After ingots pulling, the ingots will be cooling down rapidly, stress will be remained inside the ingots, and also the ingots will experience a lot of processes such as slicing, edge grinding etc. During the mechanical process, stress and defects will be imported into the wafer. Stress is easier to release and small defects are easier to move to gathering during thermal process, the higher temperature (lower than 1100℃) thermal process is, the more stress will be released and smaller defects will gather together into bigger defects. Releasing the stress inside the wafer means geometry parameters BOW and WARP will change, especially WARP data; defect gathering means we will have bigger defects inside wafers that can be observed, these bigger defects will induce some defects to epitaxial layer which can't be accepted.

We prepared 40 wafers, including different dopant and different orientation, we measured the geometry data before thermal process, then 20 went to HTO and 20 went to APCVD, after that, measured them again, the tool we use here is ADE7200. We marked the result in Figure2: Warp data before and after HTO/APCVD process.

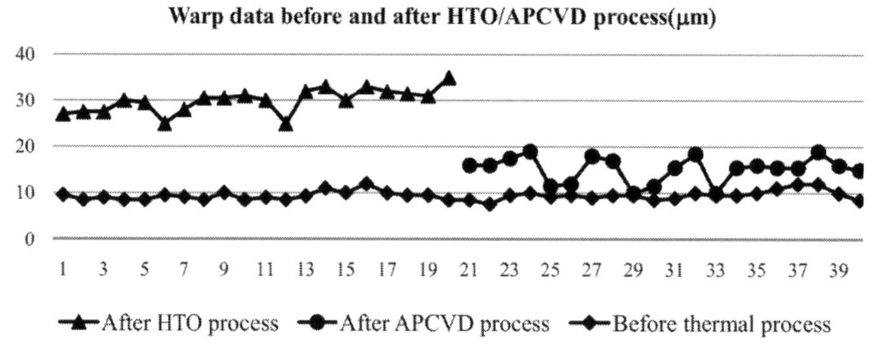

Figure2. Warp data before and after HTO/APCVD process

From the figure we can see that, before thermal process, the ◆ marks show us that all the 40 pcs wafers have quite the same WARP data, near 10 um. After thermal process we can see that, the ▲ marks and •marks have quite different data, wafers with HTO process the WARP data changed much more than wafers with APCVD process, the average WARP data is near 32 um after HTO process, and the average WARP data is near 16 after APCVD process. Now we will know that APCVD will do better job in controlling WARP data, the experiment result is in accordance with theory analysis. We didn't observe any difference between different dopant nor different orientation.

After measuring WARP data by ADE7200, we chose wafers with different orientation to observe defect, the wafers are from both APCVD and HTO. In order to make the observation easier, we polished all the wafers, and then used etching solution to etch them, and then through the microscope we would get the results. The theory is that, after polishing, the surfaces of wafers are quite smooth, but defects still exist although can't be seen. In the etching solution, the defect area will be etched much faster than normal area free from defects, in this way, after etching process, we will get different surface topography on the wafer surface. As is mentioned above, the temperature of HTO process is higher we should have more defects observed from the wafers after this process. The coincidence is that the temperature for Oxide Induced Stacking Fault (OISF) is quite the same as HTO process, it means that, we may get much more defects after HTO process if there is small defects originally exist before HTO process. Figure3: Microscope photos after HTO process show us that, some wafers will be induced defects during HTO process, but some wafers are not, depend on if the smaller defects exist inside the wafers or not. For <100> and <111> we would get quite typical defect figure after HTO, for <100> orientation, square appeared, and for <111> orientation, triangle appeared. There figures are the combination of stacking faults and dislocation, at each vertex of the figure there are dislocations, and stacking faults will connect them together to make the square and triangle. These defects can't be seen after polishing, but during epitaxial process, these hidden defects will lead to epitaxial defects such as stacking faults and slip-line which can't be accepted.

From APCVD processed wafer, we did not observed any defects, because the temperature for APCVD is only 500℃ and the process time is only 5 minutes, far from the OISF temperature. Although smaller defects exist in the wafers, they do not have enough time and energy to move to get together. So after APCVD process we get photos all like the one from HTO which is free from defects.

<111>/<100> orientation free from defects

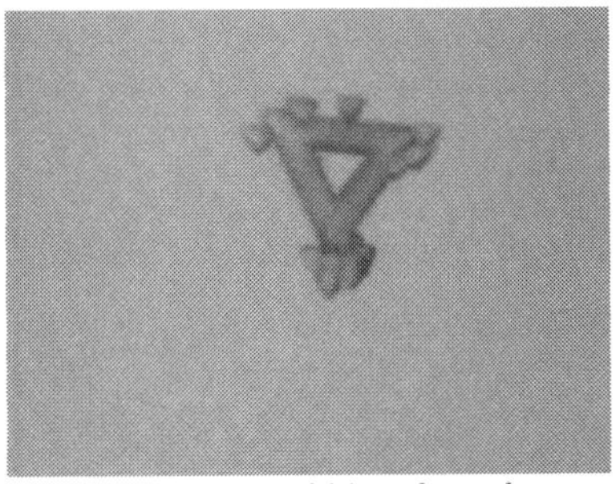

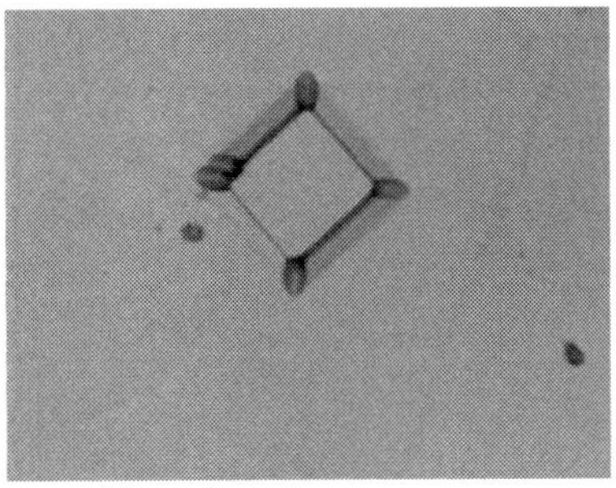

<111>orientation <100> orientation

Dislocations and stacking faults combination

Figure3. Microscope photos after polishing with HTO process

This experiment means that HTO process is easier to induce defects, because of the high temperature and long time, but if we can control the smaller defects during the process before HTO process, we can also get defect free wafers after HTO. The original of the defects after HTO we observed is the smaller defects already existed inside the wafers.

From this section we will know that, due to the high temperature and long time in HTO process, WARP will deteriorate more than APCVD which has lower temperature and quite short process time. Also, if wafers have defects before HTO, HTO process will gather them together and make the combination harmful, but APCVD will not do this also due to the lower temperature and quite short process time.

4 Conclusion

From the above experiments, we know that HTO and APCVD, each process has its own advantage and disadvantage. HTO will do better job in the controlling of thickness homogeneity and compactness of silicon dioxide layer. APCVD is good at WARP controlling and defect inducing. One more advantage for APCVD wafers is that after thermal process, we do not need to remove the silicon dioxide layer with HF before polishing, which is necessary for HTO wafers.

In summary, which process we choose depend on the result we want and the usage for the product. In order to get better WARP, less defect and not much concern about the thickness homogeneity and compactness of silicon dioxide layer, we should choose APCVD. In order to get better thickness homogeneity and compactness of silicon dioxide layer, we should choose HTO. But if we want all of the advantages, we must control the level that defect and stress induced into the wafers very low before the HTO process, or else, HTO will destroy the wafers.

References

1. Weifeng Sun, Bo Zhang, Shengan Xiao. Development and expectation of power semiconductor devices and power integration technology, *SCIENCE CHINA Information Sciences*. 42(2012): 1616-1630
2. Zhenhuai Liu, Ge Shi, Wenwei Wang, Wen Wang. TEOS LPCVD technology application in the silicon sealing technology. *New progress in materials science and engineering* (2004)
3. Yuling Liu. Research on silicon C M P polishing technology, *Electronics Process Technology*. 31,5 (2010) :299-333
4. Yunbiao Wang, Yanan Chen, Yuling Liu. Research on ultra-thin silicon film polishing technology, *Electronics Process Technology*. 32,1(2011) :48-50
5. Dan Gao, Liying Tong. Research on surface particles of LPCVD back sealing wafer, *Electronics Process Technology*. 03(2014)
6. A.C. Admas, C.D.Capio. The deposition of silicon dioxide films at reduced pressure. *Journal of the Electrochemical Society*. 126(1978): 1042-1046

DOI: 10.1051/matecconf/20178802001

A Facile and Green Synthetic Route for Preparation of Heterostructure Fe_3O_4@Au Nanocomposites

Liping Xie[1,a], Wei Qian[1,a], Sheng Yang[2], Jianjun Sun[1] and Tianxing Gong[1].

[1]*Sino-Dutch Biomedical and Information Engineering School, Northeastern University, 110004, Shenyang, China*
[2]*College of Life and Health Science, Northeastern University, 110004, Shenyang, China*

Abstract. Magnetic nanoparticles offer many exciting opportunities in biology and biomedicine, such as magnetic resonance imaging, magnetic hyperthermia therapy, biomedical diagnosis. The synthesis of multifunctional magnetic nanocomposites that possess water-solubility, magnetic properties and optical stability by a green method at room temperature in aqueous phase is still an unmet need. Here, we developed a simple and green method for preparing Fe_3O_4@Au integrated the super-paramagnetic and optical properties by seed-mediated growth at mild condition in aqueous phase. The amphiphilic, non-ionic and nontoxic polymer poly(vinylpyrrolidone) (PVP) was used as a coupling agent for synthesis of Fe_3O_4@Au nanocomposites, which avoided the direct connection of Au and Fe_3O_4, and improved the saturation magnetization values of Fe_3O_4@Au to 40 emu/g at room temperature. We anticipate that the multifunctional Fe_3O_4@Au nanocomposites with high magnetic and good optical properties will provide a platform for potential diagnostic and therapeutic biomedical applications.

1 Introduction

Magnetic iron oxide nanoparticles (MNPs) have attracted considerable interest due to its superparamagnetism, and offered many exciting opportunities in biology and biomedicine, including magnetic resonance imaging, magnetic hyperthermia therapy, biosensor, separation and purification of protein and cell [1-5]. Meanwhile, the gold NPs have been extensively used in biosensor, surface enhanced Raman spectroscopy, and therapy applications, due to its well-known catalytic activity, optical properties, chemical functionality, and biocompatibility [6-8]. Bifunctional nanomaterial integrates gold and iron oxides into a single nano-platform, and by combining the advantages of gold and iron oxides, it shows enhanced physical and chemical properties [9]. Au shell layers provide stability and easy functionalization to the nanocomposites in solution, and render the magnetic NPs with plasmonic properties.

There are a number of literatures about the synthesis of MNPs@Au nanocomposites. Various nanocomposites with high shape regularity and a narrow size distribution were prepared by the thermolysis of organometallic precursors in hot organic solutions (180–190 °C) [10,11] , or via thermally activated hetero-interparticle coalescence between gold and magnetic NPs under an

[a] Corresponding author: xielp@bmie.neu.edu.cn and weiq@bmie.neu.edu.cn

encapsulating environment at temperature elevation (140–160 °C) [12]. All these methods involve hot organic solutions, resulting in the NPs capped with organic surfactants, which render them unsuitable for biomedical applications. Sun et al. synthesized the oleylamine and oleic acid capped Fe_3O_4 NPs via thermal decomposition of iron(III) leate in a mixture of oleylamine and oleic acid, and then, they synthesized the Fe_3O_4/Au NPs in a chloroform solution in the presence of the Fe_3O_4 NPs. Core/shell Fe_3O_4/Au NPs could be used as seeds for continued growth of Au shell or Ag shell in in aqueous solution. [13] Lyon et al. provided a rapid and effective route for synthesis of magnetic core/shell particles that were soluble in aqueous media [14]. However, this method can't form Au shell on Fe_3O_4 NPs directly. The Fe_3O_4 NPs need to be exposed to air for extended periods (~1 week), or oxidized to form γ-Fe_2O_3, which decreases the saturation magnetization and elongates the synthesis time. Liu et al prepared ultra-small dimensional PVP-coated Fe_3O_4/Au nanocomposites in octyl ether by nanoemulsion polyol process [15]. However, it also needs high temperature (280 °C) to form the Fe_3O_4/Au nanocomposites, and the transfer of the NPs from organic phase to aqueous phase. Miao et al provided a facile route at room temperature in aqueous phase [16], through 11-mercap-toundecanoic acid used as a linker, for preparation of magnetically responsive Fe_3O_4@Au. Unfortunately, this method requires tedious washing and sonicating processes as well as long preparation time (over one day). Consequently, routine synthesis often involves high temperature, several separate and time-consuming processes in order to obtain such heterostructure nanocomposites. Thus, the synthesis of a uniform and monodisperse multifunctional nanocomposite that suitably integrates Fe_3O_4 NPs and gold nanoshells, and possesses water-solubility, magnetic properties and optical stability by a rapid simple methodology at room temperature in aqueous phase is still a great challenge.

Here, we developed a simple and green method for preparing Fe_3O_4@Au by the seed-mediated growth. The Fe_3O_4 particles were prepared by coprecipitation of an Fe(II) and Fe(III) salt in alkaline medium at room temperature. The amphiphilic, nonionic, nontoxic, polymer poly(vinylpyrrolidone) (PVP) was used as a mediator to stabilize Fe_3O_4 NPs. It provided sufficient steric stabilization for the colloids to avoid aggregation. On the other hand, it was small enough to form a relatively homogeneous and slightly negative charge layer onto Fe_3O_4 NPs colloids. The number of NPs decorated with PVP could be controlled by changing the molecular weight of PVP [17]. Gold seeds (~4 nm) with positive charge were absorbed on the PVP modified Fe_3O_4 NPs. Au shells were formed by reduction of Au^{3+} onto the surfaces of Fe_3O_4 NPs by the hydroxylamine hydrochloride. To the best of our knowledge, the rapid synthesis of three-layer nanocomposites with Fe_3O_4 as the core, PVP coating as a media and Au shell by room temperature in aqueous solution with high yield have not been reported before. Such Fe_3O_4@Au nanocomposites show excellent monodispersity in particle size, outstanding magnetic response, well-defined optical properties and long-term stability, which has made it promising for biomedical applications.

2 Experiment

2.1 Materials

Poly(vinylpyrrolidone) (PVP, Mw 58,000), ferrous chloride ($FeCl_2 \cdot 4H_2O$), iron(III) chloridehexahydrate ($FeCl_3 \cdot 6H_2O$), sodium hydroxide (NaOH) and hydroxylamine hydrochloride (NH_2OH_3HCl) were purchased from Aladdin ®(Shanghai, China). Tetrachloroauric acid ($HAuCl_4 \cdot 3H_2O$, 99.99%) and cetyltrimethyl ammonium bromide (CTAB) were purchased from Sigma-Aldrich (New Jersey, USA). Deionized water (DI, 18 MΩ cm^{-1}) was used in all experimental processes as needed. All materials were used as received without further processing.

2.2 Synthesis of Fe_3O_4 NPs

$FeCl_3 \cdot 6H_2O$ (1.825 g) and $FeCl_2 \cdot 4H_2O$ (0.895 g) were dissolved in 15 ml of degased DI water. 5 ml of NaOH solution (5.8 M) was added into the iron chloride mixture dropwise by a syringe. The reaction was continues for 30 minutes with continuous stirring at room temperature under nitrogen protection. Then, washed the black precipitate with magnetic separation twice. The black Fe_3O_4 NPs were re-dispersed in PVP solution with weight ratio of 1:5 to form a stable suspension.

2.3 Synthesis of Fe_3O_4@Au nanocomposites

Firstly, 4 nm gold seeds were prepared through borohydride reduction of gold salt in the presence of CTAB, which also made the gold seeds positively charged. The fresh sodium borohydride solution (0.01 M) was added to the mixture of 0.1 M of CTAB and 0.2 mM gold chloride acid solution. The color of the mixture turned from yellow to light brown, indicating gold seeds were formed. 2.8 ml of diluted Fe_3O_4 NPs solution (2mg/ml) was stirred with 0.5 ml gold seeds for 1 hours. The mixture was isolated by magnet, and the precipitates were re-dispersed in the mixture of 4 ml DI water, 0.4 ml (1% wt.) citric acid and 0.2 ml 0.025M $HAuCl_4 \cdot 3H_2O$. The pH of the mixture was adjusted to pH=9 by NaOH, then, 0.2 ml of hydroxylamine hydrochloride was added. The color of the reaction solution changed from black to red wine, indicating Fe_3O_4 @ Au nanocomposites were successfully synthesized.

3 Results and discussions

The approach for the synthesis of the Fe_3O_4 @Au nanocomposites was illustrated in scheme 1. A"seed-mediated growth"strategy was used to synthesize the Fe_3O_4 @Au nanocomposites [18]. We chose a two-step method for preparation of Fe_3O_4 @Au nanocomposites. As shown in scheme 1, the Fe_3O_4 NPs were fabricated firstly by coprecipitating of Fe^{2+} and Fe^{3+} ions (in the molar ratio of 2:3) by alkaline solution. After 30 mins' reaction, the Fe_3O_4 NPs were precipitated by magnet, and then washed and re-dispersed in DI water. After that, the amphiphilic, nonionic, nontoxic PVP was used as a mediator to stabilize Fe_3O_4 NPs. It provided sufficient steric stabilization of the colloids to ensure well-dispersed magnetic particles NPs. On the other hand, it formed relatively homogeneous and slightly negative charge layer onto Fe_3O_4. Gold seeds were synthesized by borohydride reduction of gold salt in the presence of CTAB. The surfactants CTAB not only used as stabilizer for the Gold seeds, but also made the gold seeds negatively charged. The gold seeds were absorbed on the Fe_3O_4 NPs by the electrostatic interactions between the slightly negatively charged PVP on Fe_3O_4 NPs and positively charged gold seeds. The water-soluble core/ shell Fe_3O_4/PVP/ gold seeds serve as seeds for the formation of Fe_3O_4@Au nanocomposites with thicker Au coating. Hydroxylamine hydrochloride was used as a mild reducing agent to form thicker Au coating on Fe_3O_4 by simply adding more $HAuCl_4$ in alkaline condition. The processure involved herein was perfom at room temperature and in aqueous phase. The mild and simple synthetic conditions reported here avoided the tedious procesure to transfer nanocomposites from organic phase to aqueous phase, and avoided the usage of toxic solvents.

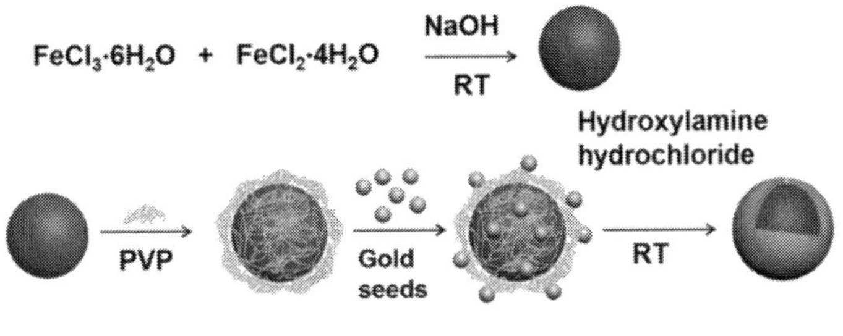

Scheme 1. Schematic synthetic route for preparation of heterostructured Fe_3O_4@Au nanocomposites at room temperature in the aqueous phase.

We further characterized the optical and magnetic properties of synthesized magnetic nanocomposites. The product of Fe_3O_4@Au was dispersed in aqueous solution with a well monodispersity. The nanocomposites exhibited red wine color, which indicating the formation of Au shell on the magnetic nanocomposites (Figure 1a). The nanocomposites were effectively separated from the solution in the presence of a magnetic field by a magnet, leaving the bulk solution clear and transparent (Fig. 1b), which demonstrated the effective magnetic separation characteristics of Fe_3O_4@Au nanocomposites. After the removal of the external magnet and with slightly shaking, the transparent solution turned to red wine color again as shown in Figure 1a. The respective absorption spectra measured for Fe_3O_4 NPs, Au seeds and Fe_3O_4@Au were shown in Figure 2. The solution of Fe_3O_4 NPs and Au seeds did not show any obvious measurable surface plasmon resonance (SPR) peak in the visible region. However, the Fe_3O_4@AuNPs showed a clear SPR peak at 540 nm, indicating the formation of Au shell on Fe_3O_4 NPs. Measurements of the SPR band of the NPs provided complementary evidence of the Fe_3O_4@Au nanocomposites.

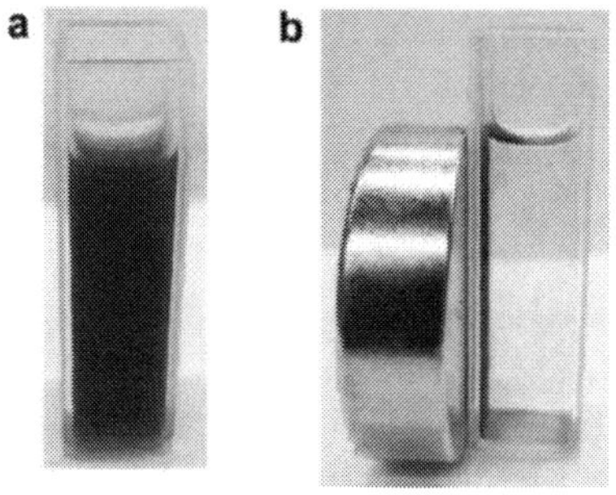

Figure 1. Photo images of Fe_3O_4@Au with absence (a) and presence (b) of an external magnet.

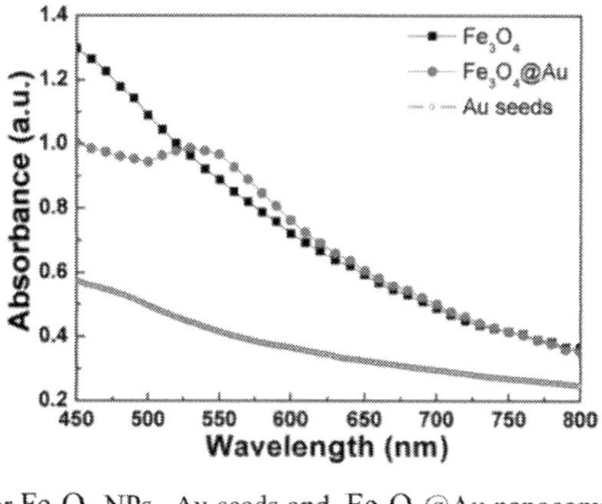

Figure 2. Absorption spectra for Fe_3O_4 NPs , Au seeds and Fe_3O_4@Au nanocomposites.

In order to further determine the crystallinity and structure of the nanocomposites, powder X-ray diffraction (XRD) spectrum was utilized. Figure 3 showed the XRD of Fe_3O_4@Au and the corresponding standard cards, JCPDS 88-0315 for Fe_3O_4 and 04-0784 for cubic gold. The line broadening observed in the XRD spectra is indicative of the small size of the NPs. All the patterns of the Fe_3O_4@Au nanocomposites are in agreement with the spectral peak of gold standard XRD diffraction spectrum (JCPDS # 4-0784), as well as the standard of Fe_3O_4 diffraction spectrum (JCPDS # 88-0315). XRD patterns for Fe_3O_4@Au NPs show characteristic peaks (at 2θ=30.157°, 35.521°, 43.172°, 53.561°, 57.098°, 62.703°), are in corresponding with their indices (220), (311), (400), (422), (511) and (440) of Fe_3O_4 with cubic phase (JCPDS 88-0315), indicating that the NPs are pure Fe_3O_4. Besides the XRD patterns belonging to Fe_3O_4 (JCPDS 88-0315), all the other diffraction peaks at 2θ=38.184°, 44.392°, 64.576°, and 77.547°, are identified as characteristic (111), (200),(220) and (311) planes of cubic Au (JCPDS 04-0784).

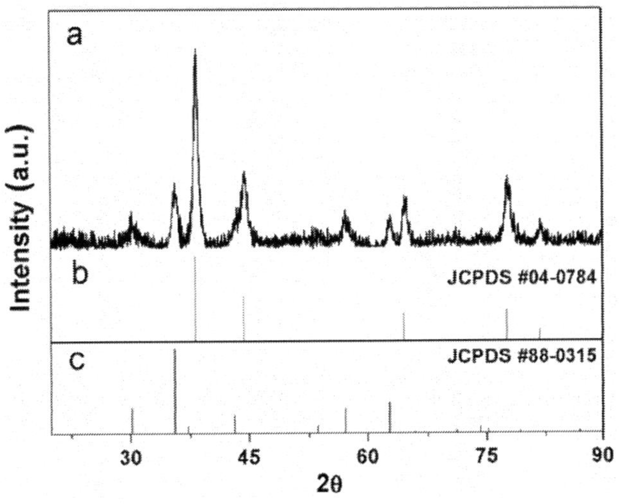

Figure 3. (a) XRD diffraction patterns of the as-prepared Fe_3O_4@Au, and the corresponding standard XRD diffraction spectrum (b) Au (JCPDS #04-0784) and (c) Fe_3O_4 (JCPDS #88-0315)

The magnetic properties of as-synthesized Fe_3O_4 and Fe_3O_4 @ Au nanocomposites were measured by vibrating sample magnetometer at room temperature (300 K). The field-dependent magnetization curves shown in Figure 4 demonstrated that Fe_3O_4 and Fe_3O_4 @ Au nanocomposites were superparamagnetic with a saturation magnetization values of 60 and 40 emu/g, respectively. A slight decline of the saturation magnetization value of Fe_3O_4@Au after coating Au was due to the diamagnetic contribution of the Au. Fortunately, the saturation magnetization value of Fe_3O_4@Au was relatively high compared with the literature reported previously [16]. The maximum magnetization value of the core/shell magnetic NPs was around 10-20 emu/g at room temperature [19]. The direct coating of Au onto magnetic NPs severely decreased the saturation magnetizationof the magnetic core by 78% or more.[20,21]. In our method, PVP as a coupling agent, not noly prevented the magnetic NPs form large-sized aggregated precipitates in solutions, but also decreased the effect of Au coating on the magnetic properties of Fe_3O_4@Au nanocomposites. As a result, the PVP-processed Fe_3O_4@Au nanocomposites with high saturation magnetisation improved the performance of operation and offer potential applications in different areas.

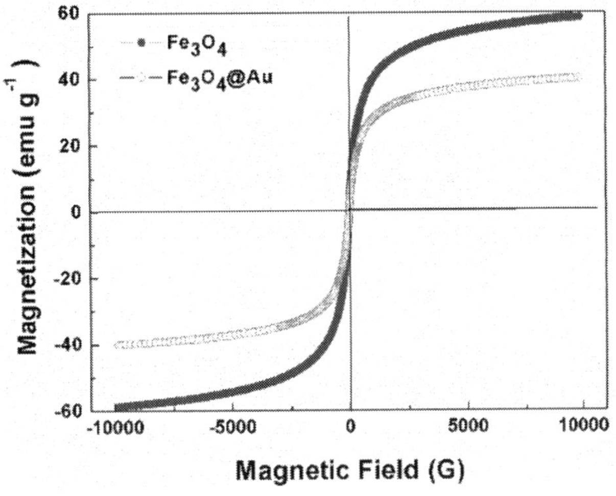

Figure 4. Hysteresis loops of (a) Fe_3O_4, (b) Fe_3O_4@Au (300 K)

4 Conclusions

In this research, we proposed a simple, rapid, non-toxic method for synthesis of water-soluble Fe_3O_4 @ Au nanocomposites in aqueous solution at room temperature. The synthesized multifunctional super-paramagnetic nanocomposites displayed interesting plasmonic properties and high saturation magnetization, which could be applied for various bio-applications. The significant contributions of the synthesis are listed as flowing. It offers a simple, fast and green method for preparing of multifunctional magnetic nanocomposites at room temperature in aqueous solution with large scale. It avoids tedious procedures to transfer the magnetic nanocomposites form organic phase to aqueous phase, and avoid the usage of toxic reagents (e.g. benzene). We anticipate that such multifunctional magnetic nanocomposites will have great potentials for nanoparticle-based diagnostic and therapeutic applications due to its magnetization and optical properties.

Acknowledgments

The authors acknowledge the financial support from the Natural Science Foundation of China (81501556, 61672146, 51502034) and the Fundamental Research Funds for the Central Universities (141903001).

References

1. C. Wen, L. Wu, Z. Zhang, Y. Liu, S. Wei, J. Hu, M. Tang, E. Sun, Y. Gong, J. Yu, D. Pang, Acs Nano **8**, 941 (2014).
2. J. Bao, W. Chen, T. Liu, Y. Zhu, P. Jin, L. Wang, J. Liu, Y. Wei, Y. Li, Acs Nano **1**, 293 (2007).
3. M. Liong, A. N. Hoang, J. Chung, N. Gural, C. B. Ford, C. Min, R. R. Shah, R. Ahmad, M. Fernandez-Suarez, S. M. Fortune, M. Toner, H. Lee, R. Weissleder, Nat. Commun. **4**, 1752 (2013).
4. Y. Pan, X. Du, F. Zhao, B. Xu, Chem. Soc. Rev. **41**, 2912 (2012).
5. Y. Yuan, Z. Ding, J. Qian, J. Zhang, J. Xu, X. Dong, T. Han, S. Ge, Y. Luo, Y. Wang, K. Zhong, G. Liang, Nano Lett. **16**, 2686 (2016).
6. K. M. Mayer, J. H. Hafner, Chem. Rev. **111**, 3828 (2011).
7. K. Saha, S. S. Agasti, C. Kim, X. Li, V. M. Rotello, Chem. Rev. **112**, 2739 (2012).
8. L. Dykman, N. Khlebtsov, Chem. Soc. Rev. **41**, 2256 (2012).
9. Y. Jin, C. Jia, S. Huang, M. O'Donnell, X. Gao, Nat. Commun. **1**, 41 (2010).
10. Q. Wei, H. Song, A. P. Leonov, J. A. Hale, D. Oh, Q. K. Ong, K. Ritchie, A. Wei, J. Am. Chem. Soc. **131**, 9728 (2009).
11. WangWang, J. Luo, Q. Fan, M. Suzuki, I. S. Suzuki, M. H. Engelhard, Y. Lin, N. Kim, J. Q. Wang, C. Zhong, J. Phys. Chem. B **109**, 21593 (2005).
12. H. Park, M. J. Schadt, Wang, I. S. Lim, P. N. Njoki, S. H. Kim, M. Jang, J. Luo, C. Zhong, Langmuir **23**, 9050 (2007).
13. Z. Xu, Y. Hou, S. Sun, J. Am. Chem. Soc. **129**, 8698 (2007).
14. J. L. Lyon, D. A. Fleming, M. B. Stone, P. Schiffer, M. E. Williams, Nano Lett. **4**, 719 (2004).
15. H. Liu, P. Hou, W. Zhang, J. Wu, Coll Surf A: Phys Eng Asp **356**, 21 (2010).
16. X. Miao, T. Wang, F. Chai, X. Zhang, C. Wang, W. Sun, Nanoscale **3**, 1189 (2011).
17. J. Cha, P. Cui, J. Lee, J. Mater. Chem. **20**, 5533 (2010).
18. K. Yong, Y. Sahoo, M. T. Swihart, P. N. Prasad, Coll Surf A: Phys Eng Asp **290**, 89 (2006).
19. C. Wu, Y. Huang, P. Chen, K. Hoshino, H. Liu, E. P. Frenkel, J. X. J. Zhang, K. V. Sokolov, Acs Nano **7**, 8816 (2013).
20. I. Chiang, D. H. Chen, Adv. Funct. Mater. **17**, 1311 (2007).
21. L. Wang, H. Park, S. I. Lim, M. J. Schadt, D. Mott, J. Luo, X. Wang, C. Zhong, J. Mater. Chem. **18**, 2629 (2008).

Preparation and Characterizations of Composite Material Based on Carbon Fiber and Two Thermoset Resins

Hany Fouda[1,a], Lin Guo[1] and Karim Elsharkawy[1]

[1]*Beihang University of Aeronautics and Astronautics, Beijing, china*

Abstract. In the present investigation, we used two types of thermoset resins (epoxy resin and phenol formaldehyde resin) with carbon fiber (CF) to produce composite materials. CF/epoxy resin composite and CF/phenolformaldhyde resin composite were fabricated and compared between their mechanical properties as compression, tension and flexural. it was found that mechanical properties of CF/epoxy composite higher than mechanical properties of CF/phenolformaldhyde resin composite such as flexural strength of CF/epoxy resin composite increased by 30 % than flexural strength of CF/phenolformaldhyde resin composite, tensile strength of CF/epoxy resin composite increased by 11.4 % than flexural strength of CF/phenolformaldhyde resin and axial compression strength of CF/epoxy resin composite increased by 14.5 % than flexural strength of CF/phenolformaldhyde resin.

1 Introduction

There is an increasing required for advanced materials with enhancement properties to meet new requirements or to replace existing materials. The carbon fibers have a unique combination of outstanding mechanical, physical and chemical properties, such as thermal resistance, high strength, and high modulus [1-3].The carbon fibers reinforced resin matrix composites is widely used in aerospace, marine, and automobile industries during the past few decades due to their good engineering properties such as lower density, high specific strength and stiffness [1-5].Resin matrix of composites based on carbon fiber determines the thermal and chemical resistance of the composite while the carbon fibers provide strength and stiffness[4].

When a composite is subjected to stress, the load is transferred from the matrix to the carbon fibers through the interface and good interfacial bonding is therefore important. It is well known that the fiber matrix adhesion strength plays an essential role on the mechanical properties of fiber reinforced polymer composites because when a load applied to composites, it will be distributed and transferred through fiber/matrix interface [6-9]. A strong bonding promotes the better involvement of more fiber, accordingly increases the strength of composites. However, carbon fibers usually perform a poor adhesion or bonding behavior to polymer matrix due to their chemical inertness and nature of smoothness [6]. In order to improve the bonding properties of carbon fibers, various approaches can be applied, which were classified into oxidative and non-oxidative treatments according to Park and Kim [10]. Composite materials are being preferred more instead of steels and other metals because of their high strength at low a specific weight and also because composites have excellent resistance to

[a] Corresponding author : hanyfouda@buaa.edu.cn

chemicals, environment and corrosion so can be used in different environments and under various conditions [1-4]. Epoxy resins are considered as high-performance thermosetting resins, which display a unique combination of properties. Epoxy resins are one of the most versatile polymers with uses across an enormously wide variety of industries. They are composed of polymeric molecules that are converted to a solid by a chemical reaction. The ability to be transformed from a low-viscosity liquid (or thermoplastic state) into a tough, hard thermoset is the most valuable single property of epoxy resins.

An epoxy system physically comprises two essential components: a resin and a curative. The curative causes the chemical reaction, which turns the epoxy resin into a solid, cross linked network of molecules. This polymer is called a thermoset polymer structure with high cohesive strength and adhesion properties because, when cured, it is irreversibly rigid and relatively unaffected by heat. However, the term epoxy can also be used to indicate an epoxy resin thermoplastic or cured state.

Epoxy resin and phenolic resin are industrially very important polymeric materials that have vast application fields. The history of the two materials started before polymer science and polymer chemistry developed. Due to the tremendous efforts of many researchers involved in the materials, resin chemistry has been elucidated and novel approaches for highly functional and high-performance materials are being developed [10]. The increases in the use of the composite materials mean that it is very important and necessary to know their properties and behaviors under working conditions. Many researches applied on CF composite materials to improve their properties as research of Asma Yasmin, Isaac M. Daniel [11], research of Yuan Xu, Suong Van Hoa [12] and research of Farhana Pervin, Yuanxin Zhou, Vijaya K.Rangari, and Shaik Jeelani [13]

Many studies about the Composites have been carried out. When epoxy resins are reinforced with high strength carbon fibers, the obtained product is used in many commercial, military and structural applications requiring low weight and high strength. They are of relatively low density and they can be formed and tailored to have stacking sequences to produce high strength and stiffness in the directions of high loading [13-14].

In this work two types of thermoset resin used to produce two different types of CF composite materials. The first prepared composite sample was CF/Epoxy resin composite and the second prepared composite sample was CF/phenolformaldhyde resin composite in order to study the effect of the bonding force of the polymeric matrix into the properties of the prepared composite as mechanical properties as tension, compression, and flexural characteristics.

2 Experimental work

2.1 Materials

The carbon fiber used is PAN carbon fiber made from a polyacrylonitrile precursor has moderate strength; moderate modulus and carbon fiber yarn contains 3000 filaments obtained from Russia. The epoxy resin used in this study is a two component epoxy system consisting of (EPON 828) cured with polyamide (versamide 125) hardener. This two component system obtained from HEXION Company, France .epoxy processed in proportions of 1:1 by weight. Carbon fiber impregnated with phenol formaldehyde resin obtained from Russia.

2.2 Preparation of epoxy samples

Epoxy resin system consists of two parts A (EPON 828) and B hardener (versamide 125), part A was carefully weighed and stochemetric amount of part B was added, then stirring using magnetic stirrer for 30 min with low velocity to prevent air trapping [13].

2.3 Composite fabrication

Composite material based on CF/epoxy resin and was prepared by hand lay-up technique. Brush and roller were used to help the impregnation of fiber. The epoxy resin was prepared first as in the previous paragraph 2.2. The epoxy was brushed on the surface of CF. the epoxy-brushed fiber tape was carefully stacked up and aligned together layer above layer then compressed adding load 50 Kg over the upper surface and left at room temp for 24 hrs [11].composite material based on CF/phenolformaldhyde resin were prepared by the same way but differ in the last step because cured at elevated temp so put in a furnace at 150 °C under pressure 70 bar for 2 hrs.

2.4 Mechanical testing

In this research, two types of mechanical testing machines were used. The first test machine is MTS810 Servo-hydraulic material test system with100 KN axial loading was applied to investigate tensile modulus and strength. The second test machine is RT30 Electro-mechanical Alliance test system with 30 KN axial loading applied to measure both flexural and compression strength .the two machine systems were obtained from MTS Company, United States.

2.4.1 compression test

The compression strength of composite specimen of dimensions [10mm ×10 mm ×20mm] was tested. The test speed was kept const at 2mm/min. Three specimens of each composition were measured and an average value was reported. In the case of composite we measured compression strength in both directions (axial and radial directions) as shown in figure 1.

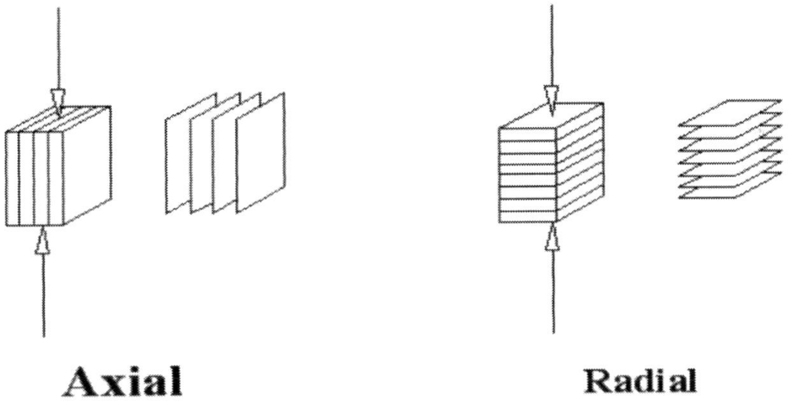

Figure 1. Axial and radial compression specimens for composite material.

2.4.2 Flexural test (the three-point bending test)

The Flexural strength is the maximum stress developed when a bar – shaped specimen is subjected as a single beam to a bending force perpendicular to the bar. The Flexural strength of composite specimen of 12.7mm width, 200mm length and thickness according to the number of layers. Three specimens of each composition were measured and an average value was reported.

2.4.3 The tensile test

The tensile strength of composite specimen of 12.7mm width, 203mm length and thickness according to the number of layers. Three specimens of each composition were measured and an average value was reported. The standard shape and dimensions for the tensile test samples and flexural test samples shown in figure 2.

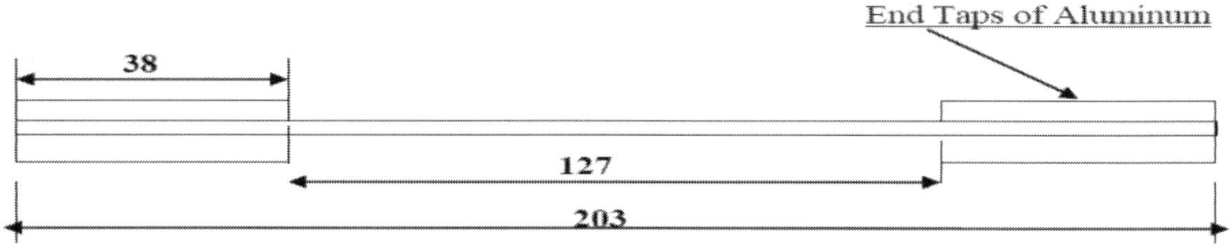

Figure 2. Standard Flat Tensile and flexural specimens.

2.5 Thermal conductivity measurements

Thermal conductivity values are used to measure heat flow through a material according to the apparatus specifications of lee ś disk method used in determination of thermal conductivity of bad conductor, the specified specimens were found to take the form of a thin disc with one side of it is heated by being placed in contact with a metallic steam chamber and the other side is in contact with a cylindrical brass calorimeter which is used for measuring the quantity of heat conducted to it per second as shown in figure 3.

Figure 3. Thermal conductivity measurement apparatus.

The test was carried out according to the following steps:
1. The specimen is placed between the steam chamber and the calorimeter.
2. Heat in the form of steam flows from the higher temperature (steam chamber) to lower one (calorimeter) for a sufficiently long time (45-60 minutes) until the thermometers give constant

reading for the temperature of steam chamber and that of the calorimeter denoted by (ϴ1 and ϴ2) respectively, then the specimen is removed.

3. The calorimeter with steam chamber on it is heated to about 5^0C above the steady temperature ϴ2, and then the steam chamber is replaced by the specimens allowing for the calorimeter to cool to a temperature lower 5 ^0C below the steady temperature ϴ2.

•Rate of heat loss by the calorimeter and specimen is calculated according to the following formula:

Q= m*s*(dϴ/dt) (1)

•Rate of conducting of heat = K*A*(ϴ1 - ϴ2)/X (2)

•Thus at steady state: K*A*(ϴ1 - ϴ2) /X = m*s* (dϴ/dt) (3)

Where:

(ϴ1 - ϴ2)/X .... Temperature gradient [^0C /m]
X................ Thickness of the specimen[m]
Q................ Rate of heat loss by the calorimeter and specimen [J/sec]
m.............. Mass of the calorimeter [Kg]
S................ Specific heat of calorimeter [J/Kg. ^0C]
A................ Area of the circular disk specimen [m^2]
K................The proportionality factor, i.e. thermal conductivity coefficient of the material [W/m.^0C].

2.6 Samples code and Compositions

The code and composition of the prepared composite samples used in this research illustrated in table 1.

Table 1. Details of composition of each formulation

Formulation code	Composition
C1	8layers[CF+ epoxy resin]
C2	8 layers[CF+ phenol formaldehyde resin]

3 Result and discussion

3.1 mechanical properties

3.1.1 Tensile test result

The tensile properties of prepared composite materials are shown in figure 4.The figure shows that there are two composite samples based on CF and have the same number of CF layers. The composite sample C1is based on CF/epoxy resin and the second composite sample C2 is based on CF/phenolformaldhyde resin. The results show that the tensile strength of composite materials based on CF/epoxy resin is higher than the tensile strength of composite materials based on CF/phenolformaldhyde resin by 11.4 % respectively. This is because the adhesion force of epoxy resin is higher than that of phenolformaldhyde resin, so the composite material based on CF/ epoxy resin matrix has higher tensile strength than that based on CF/phenolformaldhyde resin matrix.

DOI: 10.1051/matecconf/20178802002

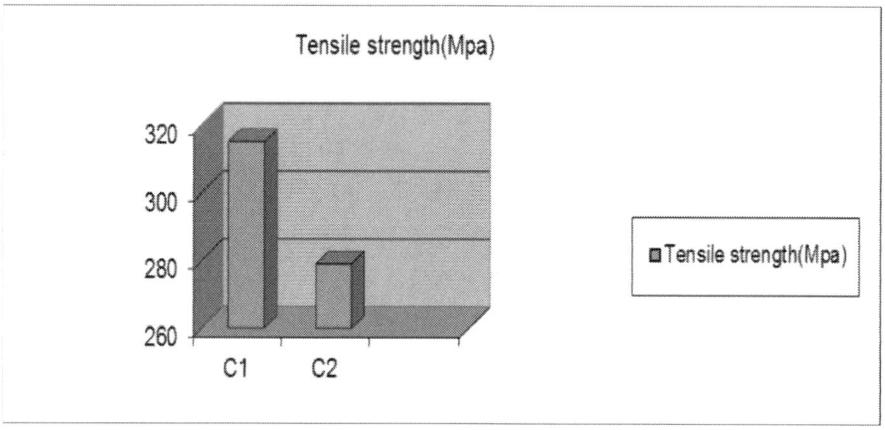

Figure 4. Tensile strength of CF/epoxy resin composite and CF/phenolformaldhyde resin composite.

3.1.2 Flexural test result

The flexural properties of the prepared composite samples are shown in figure 5. The figure shows two composite samples based on CF having the same number of layers. The first composite sample is C1 and is based on CF/epoxy resin. The second composite sample C2 is based on CF/phenolformaldhyde resin.

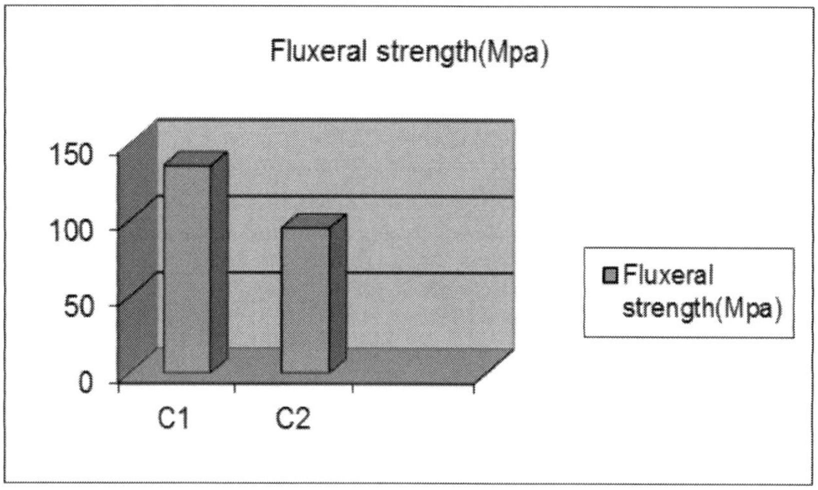

Figure 5. Flexural strength of CF/epoxy resin composite and CF/phenolformaldhyde resin composite.

From figure 5, it can be seen that the flexural strength of composite material based on CF/epoxy resin are higher than those based on CF/phenolformaldhyde resin by 30%. This is because the adhesion force of epoxy resin is higher than that of phenolformaldhyde resin.

3.1.3 Compression test result

The compression properties of the prepared composite samples are shown in figure 6 and figure 7. The figures show two composite samples based on CF and have the same number of CF layers. The first composite sample C1 is based on CF/epoxy resin and the second composite sample C2 is based on CF/phenolformaldhyde resin.

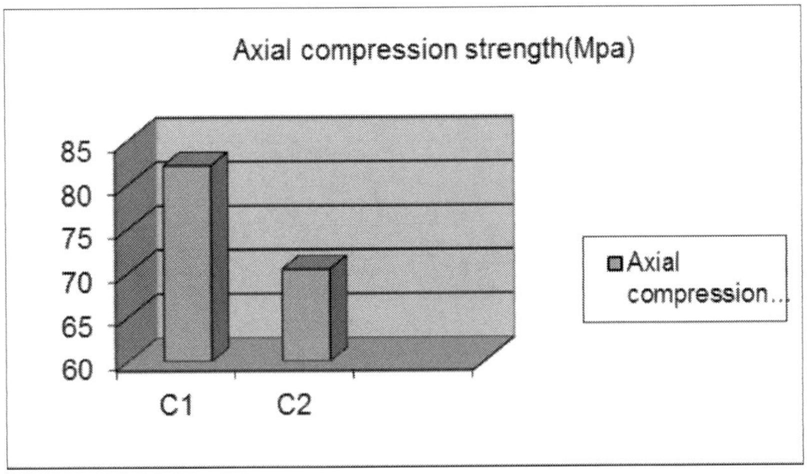

Figure 6. Axial compression strength of CF/epoxy resin composite and CF/phenolformaldhyde resin composite.

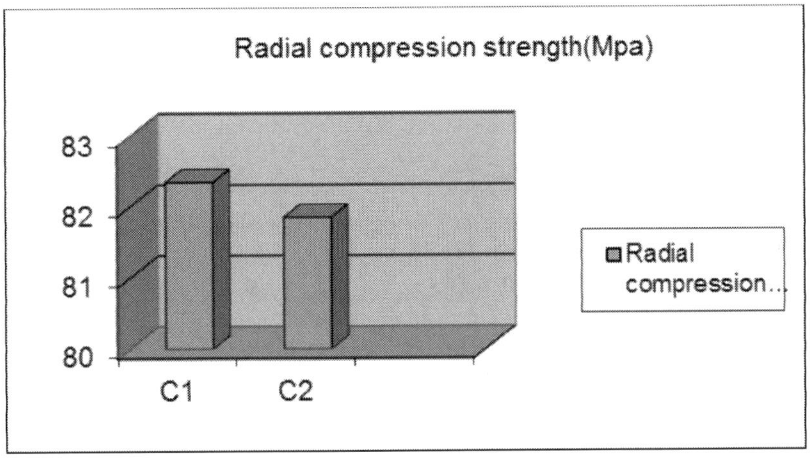

Figure 7. Radial compression strength of CF/epoxy resin composite and CF/phenolformaldhyde resin composite.

Figure 6 shows that axial compression strength of C1 is higher than that of C2 by 14.5 %. This is because in the case of axial compression the compression strength depends on adhesion force of resin. The adhesion force of epoxy resin is higher than that of phenolformaldhyde resin. Figure 7 shows that radial compression strength of C1 is nearly the same as that of C2. This is because radial compression strength depends on fiber type, not resin type which is the same in the two samples.

3.2 Thermal conductivity results

Thermal conductivity results for prepared composite material samples are shown in table 2. From the results, thermal conductivity values show that the thermal conductivity of C1<C2. The Thermal conductivity of C1 is less than that of C2.

Table 2. Thermal conductivity for different prepared composite materials.

Characteristics / Sample code	Thermal conductivity(W/m.^{0}C)
C1	0.1253
C2	0.232

4 Conclusion

Resin matrix can be used as one of the most common and effective matrix with carbon fiber to synthesis composite materials based on CF and resin matrix. It was found that as the adhesion bond of the resin matrix increase the mechanical properties of prepared composite increase. Two types of composite materials based on CF were prepared CF/Epoxy resin matrix composite and CF/phenolformaldhyde resin matrix composite. It was found that the mechanical properties of CF/epoxy resin composite higher than that of CF/phenolformaldhyde resin composite due to the higher adhesion force of epoxy resin than phenolformaldhyde resin. For example the tensile strength, flexural strength and axial compression strength of CF/epoxy resin composite is higher than that of CF/phenolformalhyde resin composite by 11.4%,30% and 14.5% respectively.it was also found that the thermal conductivity of CF/epoxy resin composite is lower than that of CF/phenolformaldhyde resin composite.

References

1. Schwartz MM," *composite materials hand book*" , McGraw-Hill; 1992.
2. Whitcomb JD, "*Composite materials testing and design*", ASTM; 1988.
3. S.T.Peters, "*Hand book of composite*", CHAPMAN&HALL; 1998.
4. Deborah D. L.Chung, "*Carbon Fiber Composites*, Butterworth-Heinemann; 1994.
5. Ronald F.Gipson,"*Principles of composite materials mechanics* ", McGraw-Hill.
6. Hui Zhang, Zhong Zang, Claudia Breidt, "*Comparison of short carbon fiber surface treatments on epoxy composites* ", institute for composite materials, university of Kaiserslautern, Erwin Schrodinger strasse 58; 2004.
7. Park BY, Kim SC, "*A study of the interlaminer fracture toughness of a carbon-fiber/epoxy composite containing surface modified short Kevlar fiber* ", Compos sci technol; 1998.
8. Fux, Chung DDL, "*Improving the bond strength between carbon fiber and cement by fiber surface treatment and polymer addition to cement mix* ", Cement concrete Res; 1995.
9. Dong S, Gauvin R," *Application of dynamic mechanical analysis for the study of the interfacial region in carbon fiber/epoxy composite materials* ", Polym compos, 1993.
10. T. Takeichi, N. Furukawa, "*Epoxy Resins and Phenol-Formaldehyde Resins* ", Reference Module in Materials Science and Materials Engineering Polymer Science: A Comprehensive Reference, pages 723-751, volume **5**, 2015.
11. Asma Yasmin, Isaac M. Daniel, "*Mechanical and thermal properties of graphite platelet/epoxy composites* ", Northwestern University, Evanston; 2004.
12. Yuan Xu, Suong Van Hoa, "*mechanical properties of carbon fiber reinforce epoxy/clay nanocomposites* ", center for research in polymer and composites, Canada; 2007.
13. Farhana Pervin, Yuanxin Zhou,Vijaya K.Rangari,Shaik Jeelani, "*Testing and evaluation on the thermal and mechanical properties of carbon nano fiber reinforced SC-15 epoxy* ", Center of advanced material, Tuskegee; 2005.
14. J.B. Donnet, compos.Sci. Technol, 63(2003) 1085-1088.

Photocatalytic Removal of Dye and Reaction Mechanism Analysis over Y_2O_3 Composite Nanomaterials

Xiaohui Guo[1,a], Jinyan Liu[1] and Guibao Guo[1]

[1]*School of Chemistry and Chemical Engineering, Inner Mongolia University of Science & Technology, Baotou 014010, Inner Mongolia, China*

Abstract. Y_2O_3 supported photocatalysts MOx (M=Fe, Ti and Bi)/Y_2O_3 were synthesized by hydrothermal and deposition-precipitation method and used in photocatalytic degradation of xylenol orange and rhodamine B under UV light irradiation. The crystalline structure and optical properties were well characterized by XRD and UV-vis DRS. The results of XRD revealed that the MOx/Y_2O_3 were composed of Fe_2O_3, TiO_2, $Bi_{0.8}Y_{1.2}O_3$ and Y_2O_3. The UV-vis DRS showed that MOx/Y_2O_3 photocatalysts exhibited stronger absorption in ultraviolet, and the absorption edge shifted to visible light region significantly. The photocatalytic experiments indicated that MOx/Y_2O_3 photocatalysts showed better activity for photodegradation of xylenol orange than rhodamine B. Moreover, the role of Y_2O_3 and the mechanism of photocatalysis is proposed.

1 Introduction

Rare earth (RE) elements are extensively exploited in metallurgical machinery, petroleum chemical industry, glass-ceramics, agriculture and functional materials due to their unique optical, electrical, magnetic and catalytic properties. However, the imbalance exploitation of RE elements severely restricted the development of RE industry, especially the overcapacity of light RE elements Ce, La and Y, so it is very important to tackle the application of light RE elements. Light RE elements as catalysts or catalytic components has been widely applied in petroleum chemical industry, catalytic combustion of fossil fuels, automotive emissions control, the purification of industrial waste air, and solid solution fuel cells [1] and showed excellent performance, that was ascribed to be one of the most effective way of application and foreground.

Photocatalysis is a promising technology for solving energy and environmental issues. In most of rare earth oxides, CeO_2 has been successfully used as photocatalysts to degrade sewage and pollutants because of its strong redox properties of Ce^{3+}/Ce^{4+} and the formation of oxygen vacancies [2, 3]. While La and Y only acted as dopant in photocatalysts, such as La/Y doped TiO_2 [4-6], Y_2O_3-ZnO [7], Y doped graphene oxide [8] and La_2O_3-Bi_2O_3 [9], had been synthesized to photocatalytic degradation of dye wastewater. The major role of RE ions in doped photocatalysts is supposed to enhance electron-hole pairs separation, improve optical absorption properties, modify adsorption capacity of dye and increase surface concentration of adsorbed OH group [10-13]. Nonetheless, the mechanism and active species of Y doped photocatalysts has not been thoroughly studied.

[a] Corresponding author : gxhsxicc@163.com

In this work, Y_2O_3 supported photocatalysts were synthesized via hydrothermal and deposition-precipitation method. XRD and UV-vis diffuse reflectance spectra were carried out to characterize the crystalline structure and optical property. The photocatalytic activity was tested by degradation of xylenol orange (XO) and rhodamine B (RhB) aqueous solutions under UV light irradiation. The photocatalytic mechanism is proposed based on the band potential analysis and active species trapping experiment.

2 Experimental

2.1 Catalyst preparation

All chemical reagents were of analytical grade and used without further purification. MO_x/Y_2O_3 supported photocatalysts were synthesized by two-step (hydrothermal and deposition-precipitation) process. Firstly, Y_2O_3 support was prepared as follows: 2 mmol $Y(NO_3)_3 \cdot 6H_2O$ was dissolved in 20 mL of deionized water, and then 60 mL KOH aqueous solution (5 mol/L) was added into the solution to form white precipitation. After being magnetically stirred for 1 h, the mixture was transferred into a 100 mL Teflon-lined autoclave which was subsequently heated at 180 °C for 12 h. After cooling down to room temperature, the solid products were filtered, washed with deionized water, dried at 60 °C overnight, calcined at 500 °C for 5 h, and finally Y_2O_3 powder was obtained.

Secondly, MO_x/Y_2O_3 photocatalysts were prepared by deposition-precipitation methods. Typically, 0.2 g of the as-prepared Y_2O_3 powder was dispersed in 20 mL of deionized water and sonicated for 30 min, and then desired amount of 20 mL of $Fe(NO_3)_3 \cdot 9H_2O$ aqueous solution was added. The suspension was magnetically stirred for 30 min and 40 mL of K_2CO_3 solution (0.5 mol/L) was added under vigorous stirring for 3 h. Subsequently, the mixtures were filtered, washed with deionized water, dried at 60 °C overnight, calcined at 500 °C for 2 h, and finally FeO_x/Y_2O_3 samples were obtained. Bi_2O_3/Y_2O_3 sample was synthesized by analogous procedure, except that $Bi(NO_3)_3 \cdot 5H_2O$ was dissolved in 5 mL of 1 mol/L HNO_3 solution. While TiO_2/Y_2O_3 sample was prepared as follow: 0.3 ml of tetrabutyl titanate was added into 80 ml of ethanol solution containing 0.2 g of Y_2O_3 powder and sonicated for 30 min to form mixture A, and then 1 ml $NH_3 \cdot H_2O$ (25%-28%) was added dropwise into A to make titanium source hydrolyze accompanying with continuously stirring at room temperature for 3 h. The resulting mixture dried by rotary evaporation at 80 °C, calcined at 500 °C for 2 h, and finally the product was denoted as TiO_2/Y_2O_3. The weight contents of MO_x in all of MO_x/Y_2O_3 photocatalysts were 25 %.

2.2 Characterization

Powder X-ray diffraction profiles were performed on Bruker D8 ADVANCE diffractometer at a scanning rate of 5°/min in the 2θ range from 20° to 80°, with Cu Kα radiation (λ=1.5406 Å) at 40 kV and 40 mA. Diffuse reflectance spectra (DRS) were carried out on Agilent Cary 100 UV-vis spectrophotometer in the range of 200-800 nm using $BaSO_4$ as reference.

2.3 Photocatalytic activity

The photocatalytic activity of all samples was evaluated toward photodegradation of xylenol orange (XO) and rhodamine B (RhB) aqueous solutions under UV light irradiation (8 W, λ=365 nm). In details, 100 mL suspension containing 0.2 g/L of catalyst and 20 mg/L of XO or RhB was sonicated for 5 min, magnetically stirred in dark for 1 h to ensure an adsorption-desorption equilibrium, and then irradiated under UV light. Simultaneously, at given intervals, small aliquots of suspension were sampled and centrifuged to remove catalysts. The XO or RhB concentrations in filtrates were analyzed on UV-vis spectrophotometer (Analytik Jena, SPECORD 50) in the range of 200-700 nm using H_2O as reference.

The process of active species trapping experiments is the same as photocatalytic tests; besides, ascorbic acid (AA, 0.1 mmol/L), triethanolamine (TEOA, 0.01 mol/L) and t-butanol (TBA, 0.01 mol/L) were added to dye solution to trap the superoxide radicals ($\cdot O_2^-$), holes (h^+) and hydroxyl radicals ($\cdot$OH), respectively.

3 Results and discussion

3.1 Characterization of catalysts

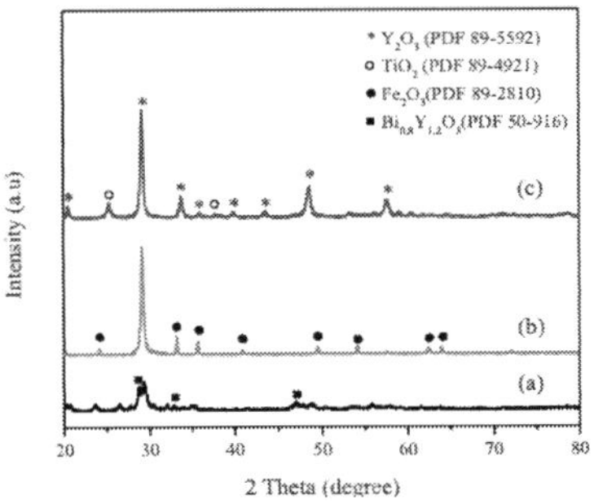

Figure 1. XRD patterns of MO_x/Y_2O_3 samples (a) Bi_2O_3/Y_2O_3, (b) Fe_2O_3/Y_2O_3, (c) TiO_2/Y_2O_3

Fig.1 shows the XRD patterns of MO_x/Y_2O_3 photocatalysts. For Bi_2O_3/Y_2O_3 (Fig.1a), two sets of XRD peaks of cubic Y_2O_3 (PDF 89-5592) and cubic $Bi_{0.8}Y_{1.2}O_3$ (PDF 50-916) were observed. Compared with Fe_2O_3/Y_2O_3 and TiO_2/Y_2O_3 samples, the intensity of diffraction peaks ascribed to Y_2O_3 dramatically decreased indicating the low degree of crystallinity, this is because Y_2O_3 can partly dissolve in HNO_3 and precipitate with Bi^{3+} together in the process of preparation. For Fe_2O_3/Y_2O_3 and TiO_2/Y_2O_3 samples (Fig.1b and c), the support YO_3 is indexed to cubic structure of Y_2O_3 (PDF 89-5592), its diffraction peak shape is irrelevant to loaded metals (Fe or Ti), and the second components can be indexed to hematite α-Fe_2O_3 (PDF 89-2810) and anatase TiO_2 (PDF 89-4921), respectively.

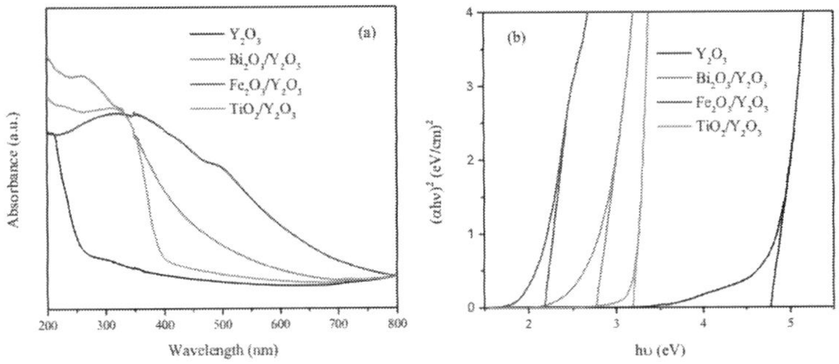

Figure 2. UV-vis absorbance spectra (a) and plot of $(\alpha h\nu)^2$ versus hν (b) of MO_x/Y_2O_3 samples

The UV-vis diffuse reflectance spectra (DRS) of Y_2O_3 and MO_x/Y_2O_3 photocatalysts were displayed in Fig.2. Taking pure Y_2O_3 support for comparison, supported MO_x/Y_2O_3 photocatalysts exhibited stronger absorption in ultraviolet, and the absorption edge shifted to visible light region significantly, especially for Fe_2O_3/Y_2O_3, which is responsive to visible light nearly to 800 nm. Moreover, Bi_2O_3/Y_2O_3 possessed strongest absorption in UV range indicating more active species ($\cdot$OH) produced on the surface of sample [7].

The band gaps (E_g) of MO_x/Y_2O_3 composites were calculated from following equation [14]: $ahv=A(hv-E_g)^{n/2}$, where α, v, E_g and A are the absorption coefficient, light frequency, band gap, and a constant, respectively; n depends on the characteristics of transition in a semiconductor, n is 4 for direct band gap and 1 for indirect band gap. Y_2O_3 is indirect band gap, so n is equal to 1 [15]. Fig.2b shows the plot of $(ahv)^2$ vs energy (hv) for MO_x/Y_2O_3 composites, the E_g can be obtained by

DOI: 10.1051/matecconf/20178802003

extrapolating the linear portion of the curve or tail. Compared to pure Y_2O_3 (4.76 eV), TiO_2/Y_2O_3, Bi_2O_3/Y_2O_3 and Fe_2O_3/Y_2O_3 samples possess E_g of 3.2, 2.8 and 2.2 eV, respectively, representing potential photocatalytic applications under visible light irradiation. The remarkable red shift of UV-vis DRS and reduction in E_g of those three supported samples were due to the fact that absorption spectral lines in Fig.2a are the superposition of each individual component. Coupling effectively promote the separation of photogenerated electrons and holes between two composite materials, but don't change their own band gaps. Consequently, the calculated band gap in this work should be attributed to TiO_2, Bi_2O_3 and Fe_2O_3, respectively, and well coincide with the values in Ref. [16].

3.2 Photocatalytic activity

Fig.3 reveals the photodegradation activity of XO and RhB in the presence of MO_x/Y_2O_3 photocatalysts, and the inset of figure is the structure of XO and RhB. For photodegradation of XO, TiO_2/Y_2O_3 and Fe_2O_3/Y_2O_3 samples showed faster photocatalytic rate in the early 90 min. As time increased further, the photodegradation rate of Bi_2O_3/Y_2O_3 sample increased rapidly and exceeded that of Fe_2O_3/Y_2O_3 sample. Fig.3a shows that the photodegradation of XO obeys the pseudo-first-order decay kinetics, the rate was illustrated as following formula: $ln(C_0/C)=k_{app}\cdot t$, where k_{app} is the apparent rate constants (min^{-1}). Plotting $ln(C_0/C)$ *vs.* t gives k_{app} from slope of linear fit, and keeps linear regression values R^2>0.97. The k_{app} were found to be 0.0102, 0.0105 and 0.0120 min^{-1} for Fe_2O_3/Y_2O_3, Bi_2O_3/Y_2O_3 and TiO_2/Y_2O_3, respectively, implying TiO_2/Y_2O_3 sample have the best photocatalytic activity.

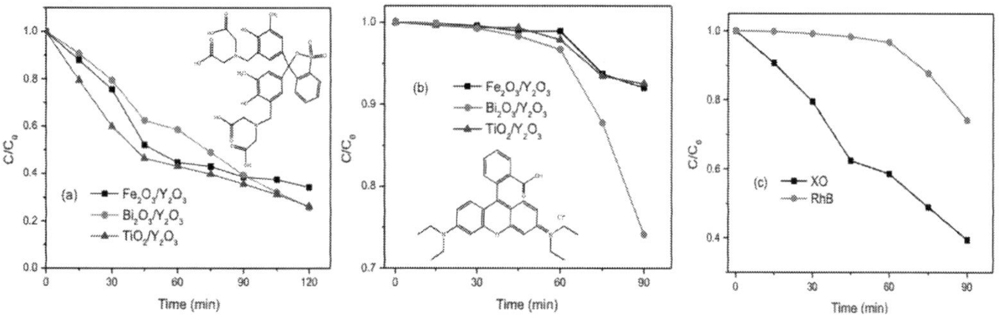

Figure 3. Photodegradation performance of (a) XO, (b) RhB over MO_x/Y_2O_3, and (c) comparison of photodegradation activity over Bi_2O_3/Y_2O_3 catalyst. (20 mg of catalyst, 100 mL of XO/RhB aqueous solution 20 mg/L)

Whereas for photodegradation of RhB, the photodegradation rate of three samples increased very slowly in 60 min, then raised approximately linearly in the last 30 min, and Bi_2O_3/Y_2O_3 sample performed the highest activity within 90 min. Nevertheless, all of three samples showed very low activity for photodegradation of RhB compared to photodegradation of XO, this was seen obviously in Fig.3c. The activity difference between XO and RhB might be derived from their molecular structure difference. From inset of Fig.3a and b, XO with four carboxylmethyl groups are more apt to be adsorbed on alkaline surface and subsequently photodegraded by photocatalysts than RhB that has only one carboxyl group. Further, the anion Cl^- in RhB is detrimental to photodegradation reaction owing to scavenging of active sites [17].

3.3 Postulated mechanism

To illustrate the photocatalytic mechanism of MO_x/Y_2O_3 and explain the origin of active species, the role of Y_2O_3 was analyzed. Y_2O_3, as a basic oxide, is abundant with surface hydroxyl and adsorbed oxygen [18], and that facilitates the photocatalytic oxidation of dye. On the other hand, coupling of band gaps between Y_2O_3 and MO_x should be taken into account.

The relative positons of valence band (VB) and conduction band (CB) edges was calculated as following [19, 20]: $E_{CB} = \chi - E_e - 0.5 \cdot E_g$ and $E_{VB} = E_{CB} + E_g$, where E_{CB} and E_{VB} are the CB and VB edge potential, χ is the absolute electronegativity, E_e is the energy of free electrons at the hydrogen scale (about 4.5 eV), E_g is the band gap. The calculated E_{VB} and E_{CB} were listed in Table 1.

Table 1. Absolute electronegativity (χ), Band gap (E_g), energy levels of calculated conduction band edge (E_{CB}) and valence band edge (E_{VB}) with respect to normal hydrogen electrode (NHE) for metal oxides.

Metal oxide	X (eV)	E_g (eV)	E_{CB} (eV)	E_{VB} (eV)
Bi_2O_3	6.23	2.8	0.33	3.13
TiO_2	5.81	3.2	-0.29	2.91
Fe_2O_3	5.88	2.2	0.28	2.48
Y_2O_3	5.35	4.76	-1.53	3.23

The case of Fe_2O_3/Y_2O_3 and TiO_2/Y_2O_3 were analogous to Bi_2O_3/Y_2O_3, so Bi_2O_3/Y_2O_3 was analyzed in detail. Bi_2O_3 was excited by UV light and generated electrons and holes pairs (Eq.(1)), but they don't transfer to Y_2O_3 owing to the mismatch of CB and VB edge potential between Y_2O_3 and Bi_2O_3. On the other side, the CB potential (0.33 eV) [16] of Bi_2O_3 is positive than $E(O_2/\cdot O_2^-)$ (-0.13 eV *vs.* NHE) [8], therefore, the electrons on the CB of Bi_2O_3 don't reduce adsorbed O_2 to $\cdot O_2^-$. The VB potential (3.13 eV) [16] of Bi_2O_3 is positive than $E(OH^-/\cdot OH)$ (1.89 eV *vs.* NHE) [8], and thus the holes on the VB of Bi_2O_3 can oxidized OH^- to $\cdot OH$.

Although the band potential of Y_2O_3 don't couple with that of Bi_2O_3, Y^{3+} was supposed to act intermediate for photogenerated electrons and holes pairs separation, where Y^{3+} acts as electron-acceptor (electron-trap) and electron-donor (hole-trap) to inhibit recombination of charge pairs and is reduced/oxidized to Y^{2+}/Y^{4+} individually (Eq.(2, 3)) [12]. Subsequently, the unstable Y^{2+} and Y^{4+} can release trapped e^-/h^+ pairs through Eq. (4, 5) and produce active species $\cdot O_2^-$ and $\cdot OH$. The $\cdot O_2^-$ may further react to from $\cdot OH$ (Eq.(6)).

$Bi_2O_3 + h\nu \rightarrow e^- + h^+$ (1)

$Y^{3+} + h^+ \rightarrow Y^{4+}$ (hole trap) (2)

$Y^{3+} + e^- \rightarrow Y^{2+}$ (electron trap) (3)

$Y^{2+} + O_2(ads) \rightarrow Y^{3+} + \cdot O_2^-$ (4)

$Y^{4+} + OH^-(ads) \rightarrow Y^{3+} + \cdot OH(ads)$ (5)

$\cdot O_2^- + e^- + 2H^+ \rightarrow H_2O_2 \rightarrow \cdot OH + OH^-$ (6)

$XO + (\cdot O_2^-, \cdot OH) \rightarrow$ Degradation products (7)

There are many works explored the photocatalytic active species on various photocatalysts. Chen [21] reported that $\cdot OH$ was the main active species for RhB degradation on flower-like $Bi_2O_2CO_3$, while Peng [22] suggested that h^+ and $\cdot O^{2-}$ are the main active species in RhB photodecomposition under solar light irradiation on WO_3-Bi_2WO_6.

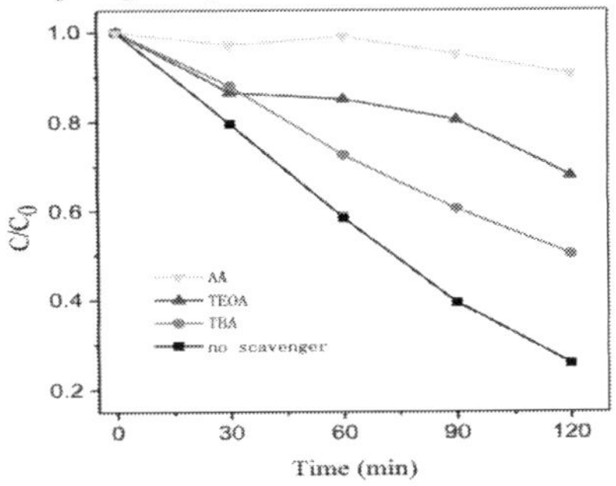

Figure 4. Effect of scavenger on photodegradation performance of XO aqueous solution over Bi_2O_3/Y_2O_3.

(20 mg catalyst, 20 mg/L of XO aqueous solution, 100 mL, 120 min)

In order to discriminate the active species in this work, the trapping experiments were carried out as before mentioned. Fig.4 shows the impact of different scavengers on the photodegradation efficiency over Bi_2O_3/Y_2O_3 sample. It is observed that the addition of AA dramatically decreased photodegradation activity, the degradation efficiency of XO declined from 74.1 % to 9.5 % (120 min). The addition of TEOA and TBA reduced degradation efficiency of XO to 32.1% and 49.8 %, respectively. These results suggest that $\cdot O_2^-$ and $\cdot OH$ are both important for photodegradation of XO over Bi_2O_3/Y_2O_3 sample under UV light irradiation (Eq.(7)).

4 Conclusions

In summary, MO_x (M=Fe, Ti and Bi)/Y_2O_3 supported photocatalysts have been successfully synthesized by hydrothermal and deposition-precipitation method. They exhibited stronger absorption in ultraviolet, the absorption edge shifted to visible light region significantly, and E_g was largely reduced with respect to pure Y_2O_3. TiO_2/Y_2O_3 and Bi_2O_3/Y_2O_3 samples showed the highest activity for photodegradation of XO and RhB, respectively. The major role of Y_2O_3 is to enhance the OH^- concentration on the surface of photocatalysts, and Y^{3+} performed synergistic effects to trap charge pairs and inhibit e^-/h^+ recombination. The $\cdot O_2^-$ and $\cdot OH$ are the main active species for photocatalytic degradation of XO over Bi_2O_3/Y_2O_3 sample.

Acknowledgments

The authors are grateful for the financial support provided by the Natural Science Foundation of Inner Mongolia of China (Grant No. 2014BS0204), Scientific Research Program in University of Inner Mongolia of China (Grant No. NJZY14163).

References

1. W. C. Zhan, Y. Guo, X. Q. Gong, Y. C. Guo, Y. Q. Wang, G. Z. Lu, Chin. J. Catal. 35, 1238 (2014)
2. N. S. Arul, D. Mangalaraj, T. W. Kim, P. C. Chen, N. Ponpandian, P. Meena, Y. Masuda, J. Mater Sci: Mater Electron. 24, 1614 (2013)
3. T. Feng, X. D. Wang, G. S. Feng, Mater. Lett. 100, 36 (2013)
4. Y. J. Wang, L. J. Jiang, C. G. Feng, Desalin. Water Treat. 52, 4802 (2014)
5. M. Meksi, G. Berhault, C. Guillard, H. Kochkar, Catal. Comm. 61, 107 (2015)
6. D. M. Tobaldi, R. C. Pullar, A. S. Škapin, M. P. Seabra, J. A. Labrincha, Mater. Res. Bull. 50, 183 (2014)
7. T. M. Su, Z. Z. Qin, H. B. Ji, Y. X. Jiang, Int. J. Photoenergy. doi:10.1155/2014/794057 (2014)
8. Y. Zhang, S. S. Yuan, Y. H. Zhao, H. G. Wang, C. D. He, J. Mater. Chem. A. 2, 7897 (2014)
9. H. Q. Deng, T. Tang, C. H. Fu, Q. Y. Jiang, Y. M. Hu, J. C. Huo, Environ. Sci. Technol. 36, 121 (2013)
10. S. Z. Chen, S. M. Xu, G. Xu, L. Y. Li, Rare Metal Mat. Eng. 35, 505 (2006)
11. L. Y. Lin, M. H. Yeh, C. Y. Chen, R. Vittal, K. C. Ho, J. Mater. Chem. A. 2, 8281 (2014)
12. A. N. Ökte, Ö. Yilmaz, Appl. Catal. B: Environ. 85, 92 (2008)
13. C. J. Hao, J. Li, Z. L. Zhang, Y. J. Ji, H. H. Zhan, F. X. Xiao, D. Wang, B. Liu, F. B. Su, Appl. Surf. Sci. 331, 17 (2015)
14. C. Saranyoo, I. Burapat, W. Khatcharin, Mater. Res. Bull. 54, 28 (2014)
15. P. K. Patel, K. L. Yadav, Physica B. 442, 39 (2014)
16. Y. Xu, M. A. A. Schoonen, Am. Mineral. 85, 543 (2000)
17. D. Kanakaraju, C. A. Motti, B. D. Glass, M. Oelgemöller, Chemosphere. 139, 579 (2015)
18. L. H. Zhao, X. M. Zheng, J. H. Fei, Chin. J. Catal. 17, 227 (1996)
19. M. A. Ahmed, E. E. El-Katori, Z. H. Gharni, J. Alloys Compd. 553, 19 (2013)
20. H. Tian, F. Teng, J. Xu, S. Q. Lou, N. Li, Y. X. Zhao, M. D. Chen, Sci. Rep. 5, 7770 (2015)
21. L. Chen, R. Huanag, S. F. Yin, S. L. Luo, C. T. Au, Chem. Eng. J. 193-194, 123 (2012)
22. Y. Peng, Q. G. Chen, D. Wang, H. Y. Zhou, A. W. Xu, Cryst. Eng. Comm. 17, 569 (2015)

DOI: 10.1051/matecconf/20178802004

Study on the Cure Kinetics of Epoxy Resin Prepreg in Fiber Metal Laminates

Xintong Wu[1], Lihua Zhan[1,a], Shujian Li[2] and Wenke Li[2]

[1]*College of Mechanical and Electrical Engineering, Central South University, Changsha 410083, China*
[2]*Light Alloy Research Institute, Central South University, Changsha 410083, China*

Abstract. To better analysis the curing behavior of a new kind of high temperature curing epoxy resin prepreg in fiber metal laminates (FMLs), dynamic differential scanning calorimetry (DSC) method was used to make the research on the curing process of the prepreg. The characteristic temperature of curing reaction and curing kinetics parameters of epoxy resin prepreg were systematically researched, and the cure kinetics equation was built. The cure kinetics model of the epoxy resin in the prepreg was determined by autocatalytic reaction model. The results showed that the reaction activation energy of the epoxy resin was 62.178kJ/mol and the reaction orders were 0.314 and 1.2157, the total reaction order was 1.5297. The optimum curing process was set at 180℃ under 0.6Mpa pressure for two hours, and the interlaminar shear strength (ILSS) of FMLs result was 69.75Mpa. The model can well describe the curing behavior of the prepreg, which will provide theoretic basis for the further studies on the thermodynamic properties of epoxy resin prepreg and the quality control of fiber metal laminates during the curing process.

1 Introduction

Fiber metal laminates are new hybrid composite structure materials which consist of the metal sheet and fiber reinforced polymeric materials, forming under a certain temperature and pressure according the needs [1]. FMLs combines the advantages of the metal and fiber reinforced resin composite. Compared with the traditional materials, FMLs have excellent fatigue and corrosion characteristics and impact resistance [2]. Due to their advantages, FMLs are widely used in most commonly in aerospace applications, now being used as structural materials for the wing skin panels, fuselage panels, fin box and rudder box. These material systems are the typical representative of advanced aerospace composites in the future.

The T800/epoxy resin used in the FMLs is a novel advanced composite materials with high toughness which developed by the Cytec company [3, 4]. Compared with traditional carbon fiber reinforced resin matrix composites, its application range is expanded to the primary structures, which can obviously reduce the weight of aircraft structure and improve the economic efficiency.

The quantitative analysis research of curing kinetics based on the change of functional groups in the reaction or the change of thermal effect. The most frequently used methods are fourier transform infrared (FTIR) spectroscopy, nuclear magnetic resonance (NMR) and DSC [5]. DSC method can directly measure the temperature change and observe the whole process than other methods. The DSC

[a] Corresponding author: yjs-cast@csu.edu.cn

method can be divided into the non-isothermal method and isothermal method. Compared to the isothermal method, the non-isothermal method can not only avoids these problems, but also has the advantages of using less sample, saving time and getting more information.

Previous researches focused on the cure kinetics of pure epoxy resin. Using isothermal and non-isothermal DSC methods, Liu[6] studied the cure kinetics of 3234 epoxy resin system. However, Guo[7] et al. found that the cure kinetics of fiber reinforced composite prepreg was definitely different from the pure epoxy resin. Although the fiber did not change on the cure reaction mechanism, the fiber in the prepreg restrained ability of molecular during the curing process. Usually, the prepreg are directly applied for the FMLs. Up to now, the researches mainly focused on the processability and service performance of the T800/epoxy resin composite, ignoring the cure kinetics and quality control during the curing process [8, 9]. Studies on the cure kinetics of T800/epoxy resin in FMLs are rarely relevant reported at home and abroad.

In this paper, the cure kinetics of T800/epoxy resin prepreg was systematically investigated through non-isothermal DSC method, getting the cure kinetics equation and the curing related parameters. The results will find out the thermodynamic properties and provide the basic theory for the further investigation of FMLs during the forming process, expanding the application range of FMLs.

2 Experimental

Carbon fiber/epoxy resin prepreg was provided by the Shenyang Aircraft Company. The reinforcement body was T800 carbon fiber and the initial fiber volume fraction was 65%. Aluminum alloys 2524-T3 sheets were supplied by Southwest Aluminum Company.

FMLs consists of thin layers of aluminum and carbon fiber/epoxy prepreg. It was 3/2 lay-up with three layers of aluminum and two layers of prepreg, bonding with autoclave.

Thermal analyzer STA449C, Netzsch. Universal material testing machine CTM5000, SUST. Optical microscope DSX 500, Olympus.

Weighing 15-20mg sample seals puts in the aluminum crucible every time. Scanning samples uses the different heating rates of 2, 4, 6, 8 and 10K/min, respectively. Before each experiment, using standard samples of indium and zinc to calibrate instrument, setting scanning temperature range from 293.15K to 573.15K.The input rate of nitrogen with high purity is 100mL/min, and recording the DSC curves.

The ILSS was done according to ASTM D2344. There were five specimens for each experiment condition and the dimensions of the sample were 12.0mm × 5.0mm × 2.0mm (length × width × thickness). The testing machine using a speed at 1.0mm/min.

3 Results and discussion

3.1 Analysis of the cure kinetics DSC curves and characteristic temperature of the curing process

Fig.1 (a) shows the T800/epoxy resin prepreg DSC curves at different heating rates. We can find that there is only one single exothermic peak in each of the heating rate during the curing process. With the increase of heating rate, the exothermic peak gradually shifts to right, the peak temperature increases and the peak becomes steep. The main reason is that the samples have enough time for curing at low heating rates. As the heating rate increases, the thermal inertia and the temperature difference also increase.

The degree of cure and curing rate with temperature changes at different heating rates were shown in Fig.1 (b). The figures show that the degree of cure with temperature changes trend is the same as the curing rate. In the lower temperature, the curing rate is slow because of the induction period of reaction. After a period of time, the temperature increases with reaction, and then the curing rate reaches maximum value. At last the curing rate decreases gradually. The curing degree and temperature change as "S" curve during whole process.

Analyzing the DSC curves at different heating rates in Fig.1 (b), we can see the characteristic temperature of the initiative curing temperature T_i, the peak top temperature T_p and the finish curing reaction temperature T_f. They were listed in Table 1.

Using linear fitting method between the different heating rates of T_i, T_p, T_f and the heating rate β deals the experimental data. When the heating rate is zero, we can calculate the isothermal curing temperature through epitaxial method. According to the linear regression equation, the prepreg curing reaction temperature T_{i0}, T_{p0} and T_{f0} is 416.12K, 455.77K and 511.21K, respectively. That is the theoretical gel temperature T_{gel}, theoretical cure temperature T_{cure} and post treatment temperature T_{treat}.

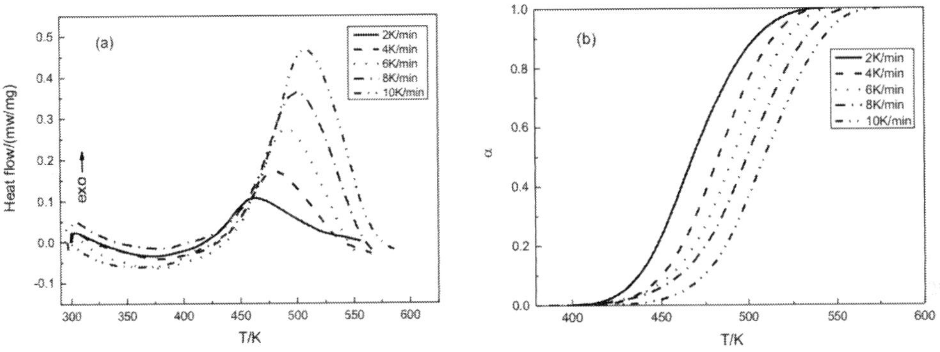

Figure 1. (a)Dynamic DSC curves of T800/epoxy resin prepreg at different heating rate(b)The degree of cure-temperature curves at different heating rate

Table 1. Characteristic temperature of T800/ epoxy resin prepreg at different heating rate

Characteristic temperature	2K/min	4K/min	6K/min	8K/min	10K/min
T_i/K	421.62	436.12	445.79	451.38	458.15
T_p/K	462.79	480.37	492.28	500.07	507.74
T_f/K	519.28	534.52	541.06	553.27	560.9

3.2 The cure kinetics model

In this paper, the phenomenological model was used to describe the curing process of T800/epoxy resin prepreg. The relationship between the curing degree and time, temperature can be clearly described by setting up the cure kinetics equation during curing process.

The autocatalytic reaction model belongs to the phenomenological model:

$$\frac{d\alpha}{dt} = k(T)\alpha^m(1-\alpha)^n \quad (1)$$

The temperature dependence of the reaction rate is described by the Arrhenius equation:

$$k(T) = A\exp(-\frac{E}{RT}) \quad (2)$$

Where "A" is the rate constant, "E" is the activation energy, "R" is the gas constant.

Fig.2 shows that the cure reaction model of T800/epoxy resin met the autocatalytic reaction model. It can be described by the autocatalytic reaction model.

3.3 The cure kinetics parameters and curing model

The activation energy of curing reaction can be obtained by Kissinger equation as follows:

$$\ln\frac{\beta}{T_p^{\,2}}=\ln(\frac{AR}{E})-\frac{E}{RT_p} \tag{3}$$

In the Eq.3, "β" is the heating rate, "T_p" is the peak top temperature, "A" is the rate constant and "R" is the gas constant. The curve of $\ln(T_p^{\,2})$-$1/T_p$ was drawn according to the Eq.3 and shown in the Fig.2. Fig.2 shows that the correlation coefficient is R^2=0.999 which indicates that the fitting precision is very high. According to its slope, we can get E=62178.45J/mol.

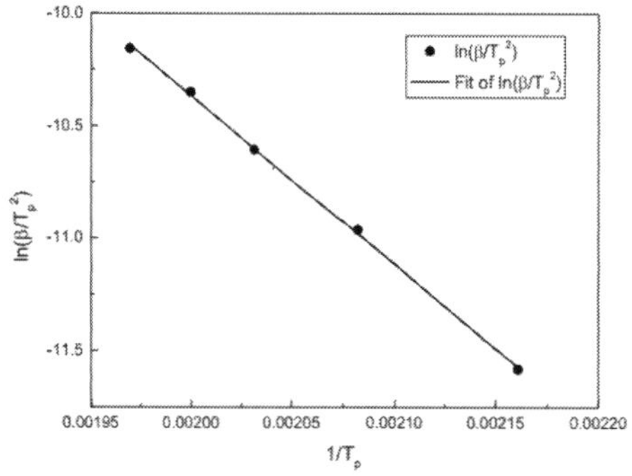

Figure 2. $\ln(\beta/T_p^{\,2})$-$1/T_p$ fitting result of prepreg's curing process

The curing kinetics parameters of T800/epoxy resin prepreg at different heating rate are shown in Table 2. According to the autocatalytic reaction model and other known parameters, we can get the cure kinetics equation of T800/epoxy resin prepreg as follows:

$$\frac{d\alpha}{dt}=17392.14\exp(-\frac{62178.45}{RT})\alpha^{0.314}(1-\alpha)^{1.2157} \tag{4}$$

Table 2. The curing kinetics parameters of T800/epoxy resin prepreg at different heating rate

β(K/min)	A(s^{-1})	m	n	R^2
2	17320.95	0.35066	1.63896	0.9976
4	17646.67	0.32288	1.30196	0.9973
6	16072.09	0.27509	1.06857	0.9886
8	14531.83	0.21784	1.01624	0.9851
10	21389.14	0.40377	1.34977	0.9928
average	17392.14	0.314	1.2157	

3.4 The curing process and ILSS results

During the FMLs forming process, the point is to control the temperature of prepreg. Appropriate temperature can directly affect the flowing ability of resin, and ensuring the interface of metal and fiber are soaked with resin. In order to improve the performance of the FMLs, the appropriate curing temperature and time are needed to cure the resin completely.

According to the cure kinetics equation, the curing cycle were done at a heating rate at 1.5℃/min up to 150℃ or 180℃ under 0.6Mpa pressure for two hours. The ILSS test results was shown in the Fig.3. Fig.4 presents the interface of FMLs which final temperature was held at 180℃.

The maximum value is 69.75Mpa and 62.56Mpa. The stress is 11 percent over the other when the curing temperature is 180℃. Because the soakage, fluidity and viscosity are better than another. There are less voids and bubbles in the interface of metal and fiber, as we can see in Fig.4. The higher bonding strength of resin with metal and fiber, the better the mechanical capabilities could be gained of FMLs.

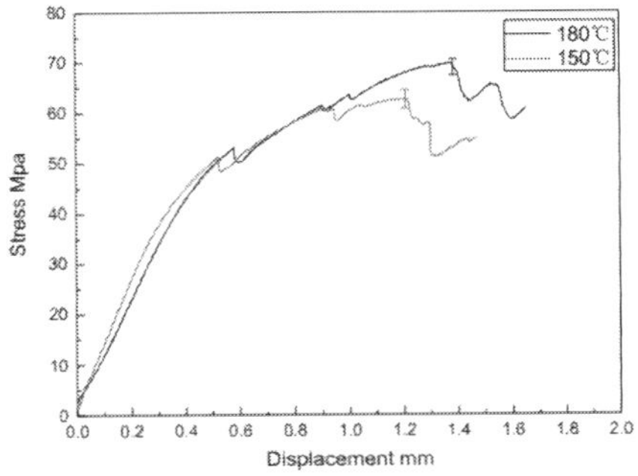

Figure 3. The ILSS test results

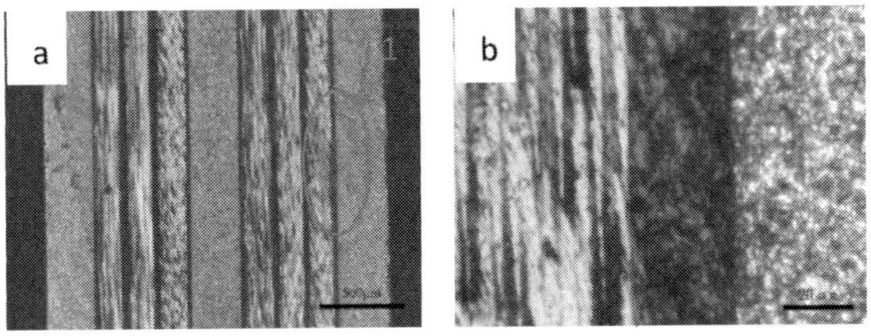

Figure 4. (a) The interface of FMLs optical image, (b) Larger image of area 1

Based on the ILSS tests, It's better to take the curing process with a temperature growth rate of 1.5℃ /min up to 180℃ under 0.6Mpa pressure for two hours. The cure kinetics equation of T800/epoxy resin prepreg provides theoretic basis for the manufacturing FMLs.

4 Conclusion

(1) In this paper, the curing reaction rate with temperature change was obtained by analyzing the DSC curves of the T800/epoxy resin prepreg. The results showed that the cure kinetics model of the T800/epoxy resin prepreg was accordant with the autocatalytic reaction model.
(2) By non-iso DSC experiment of T800/epoxy resin prepreg, we could get an equation which expressing cure kinetics of resin as: $\frac{d\alpha}{dt}=17392.14\exp(-\frac{62178.45}{RT})\alpha^{0.314}(1-\alpha)^{1.2157}$.The results of the model agreed well with the experiment results.
(3) Through the cure kinetics of the T800/epoxy resin prepreg, the curing temperature was set to 180℃ under 0.6Mpa pressure for two hours. The ILSS results was 69.75Mpa. This outcome will provide the theoretical basis for the selection and optimization of the FMLs forming.

Acknowledgements

This study was supported by the National Natural Science Foundation of China (No.51675538) and the National Program on Key Basic Research Project (2014CB046502).

References

1. T. Sinmazçelik, E. Avcu, M.Ö. Bora, O. Çoban, Materials & Design, 32, 7(2011):3671-3685.
2. L.B. Vogelesang, A. Vlot, Journal of Materials Processing Technology, 103, 1(2000):1-5.
3. C.A.J.R. Vermeeren, Applied Composite Materials, 10, 4(2003):189-205.
4. Y. Zhang, L.Zhan, Fiber Reinforced Plastics/Composites, 1, (2016):86-91.
5. F. Li, S. Zhou, B. You, L. Wu, Journal of Applied Polymer Science, 99, 6(2006):3281-3287.
6. T. Liu, B.Zhang, X.Chen, Journal of Aeronautical Materials, 1,(2005):1041-1044.
7. Z. Guo, S. Du, B.Zhang, X. Chen, Chinese Journal of Chemical Physics, 17,2,(2004):219-224.
8. X. Li, W. Gao, W. Liu, Composite Structures, 128,(2015):313-321.
9. Q. An, W. Ming, X. Cai, M. Chen, Composite Structures, 131,(2015):374-383.

Experimental Study on Tensile Properties of GFRP Bars Embedded in Concrete Beams with Working Cracks

Li Dai[1,a] and Xiongjun He[1]

[1]*School of transportation, Wuhan University of Technology, 430063 Wuhan, China*

Abstract: This paper presents the test results of an experimental study carried out to investigate the tensile properties of GFRP bars embedded in concrete beams with working cracks. The specimens were conditioned with sustained loading in 50℃ alkaline solution and tap water for 6, 12, 18 months. The tensile test results show that the degradation rate of GFRP bars embedded in the concrete specimens with work cracks is larger than that of non-work cracks, while the effect of working cracks on the elastic modulus is not significant. The microstructure of GFRP bar surface before and after the test was observed by scanning electron microscopy (SEM), combined with Fourier-transform infrared spectroscopy (FTIR) and differential-scanning calorimetry (DSC), the degree of hydrolysis reaction and glass transition temperature is also analyzed. Compared with the hydrolysis of resin matrix, it can be found that the deterioration of glass fiber and the delamination at the interfacial is more pronounced for the GFRP bars embedded in concrete beams. As a result, the mainly reason that caused the degradation of GFRP bars embedded in concrete environments are the deterioration of glass fiber and the delamination at the interfacial.

1 Introduction

In recent decades, glass fibre reinforced polymer (GFRP) bars have been used in several concrete structures all over the world due to their several advantages such as corrosion resistance, light-weight, and high strength, and it has been promoted as a solution to the deterioration of bridges, buildings and other structures made with concrete reinforced with traditional materials, such as steel. As the application of GFRP bar used in civil engineering were more and more widely[1][2].

Numerous research studies have been conducted to investigate the tensile property of GFRP bars under simulate alkali solution environment [3][4]. Chen[5] conducted accelerated tests on GFRP bars in an alkaline solution at a high temperature (60℃) for periods of 70 days. The tensile tests resulted in residual strength of 71.0% compared to the original tensile strength. Li[6]et al investigated the durability of GFRP bars exposed to harsh environment for 60 days, the environments was mixed with $Ca(OH)_2$, NaOH and KOH solution to simulate the concrete pore solution. The results showed that the tensile strength losses 11.9% of the initial strength. Wang[7] also conducted an accelerated test on GFRP bars, the reduction in the tensile strength after 183 days in the alkaline solution at 40℃ and 60℃ were 35.0% and 48.8%, respectively.

[a] Corresponding author : dlwhut2012@163.com

Thus, based on the above review, it can be found that there has achieved an abundant results. However, these studies were all focused on evaluating tensile strength of GFRP bars under artificial simulated concrete pore solution, which is different from the actual concrete alkaline environmental condition. Besides, some scholars pointed out that the traditional accelerated aging in alkaline solution is too severe and prematurely degrades the GFRP bars [8-12]. The paper presents the actual concrete environment influence on tensile properties of GFRP bars.

2 Experimental program

2.1 Test specimens

2.1.1 Raw materials

The GFRP bar used in the study was composed of 72% glass fiber and 28% vinyl ester resin, and manufactured using a pultrusion process by Nanjing Feng Hui Composite Materials Co., LTD., China. The surface of the GFRP bar was ribbed with a nylon laminate during the pultrusion. The mechanical and physical properties of 10 mm diameter GFRP bars, as provided by the manufacturer, are summarized in Table 1. The bar's circular cross section is manufactured by a process that couples pultrusion with sand coating along the external surface of the bar.

Table 1. Mechanical properties of the GFRP bar

Diameter (mm)	Ultimate tension (kN)	Tensile strength (MPa)	Elastic Modulus (GPa)
10	96	1223	51

2.1.2 Test specimens

In order to investigate the tensile properties of GFRP bars embedded in actual concrete beams, the concrete specimens were rectangular plain concrete beams with a length of 1100mm, width of 80mm, and depth of 110mm (as shown in Fig. 1), with a single longitudinal GFRP reinforcement bar centered at 30mm from the bottom of the section, and no shear reinforcement was used in the beams. Concrete beams were cast according to code for design of concrete structures (GB50010-2010), with 28 days target strength of 30 MPa. The beams were cast together with 3 concrete cubes used to determine the mechanical properties of the concrete. The concrete used for casting the beams was provided by a local ready-mix supplier with a maximum aggregate size of 10mm, and the water cement ratio is 0.4.

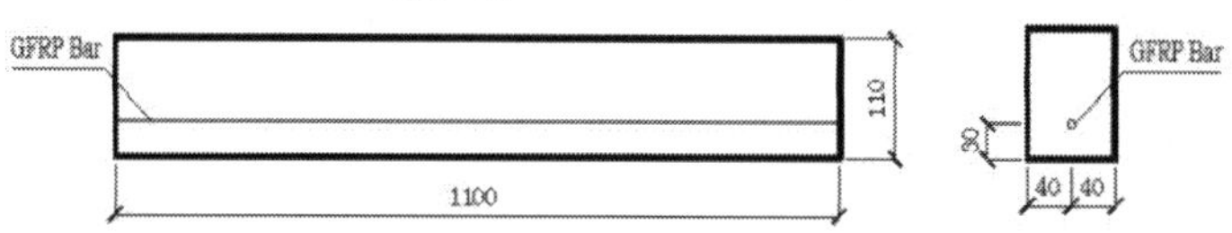

Figure 1. Geometry size of concrete beam

2.2 Test plan and environmental conditions

A spring-bracket assembly was fabricated by the research team to apply the sustained flexural loading and to produce a stable crack[13], the assembly consisted of steel frames, long threaded bolts with nuts and steel springs with a spring coefficient is 40N/mm. The sketch illustrating of the sustained loading setup is shown in Fig. 2.

 DOI: 10.1051/matecconf/20178802005

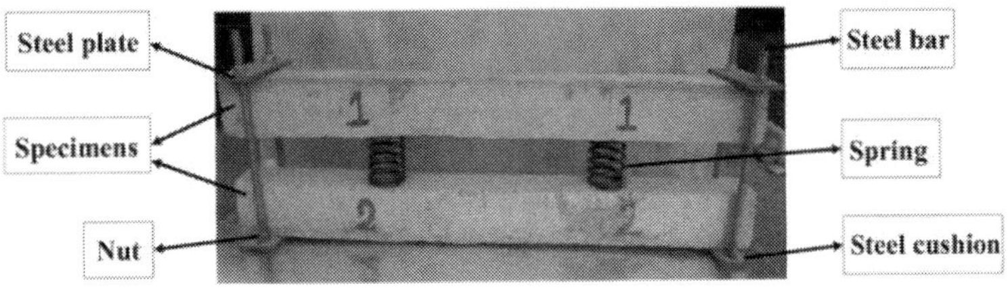

Figure 2. View of load setup

The beams were placed back to back, the inner two load points were located 200mm from the outer load points, while the two outer load points were located 100mm inward from the ends of the beam, Preliminary four-point flexure tests were performed to determine the ultimate bending moment of the beams and provide the original design data for pre-cracked load. Flexural cracks were introduced in the beams by subjecting them to the pre-cracked load. Failure of four-point flexure tests occurred in the shear/compression mode at one of the inner load points at a mean applied moment of 0.8kN·m. Beams were pre-cracked at a bending moment of 0.39kN·m. A pair of beams was not pre-cracked as a reference, and the rest of beams were pre-cracked by four-point flexure tests after a specified time.

To simulate the working state under the sustained loading, sustained four point flexural loads corresponding to 25% of the measured ultimate strength of GFRP bars were applied by spring-bracket assembly. The sustained load should be avoid concrete continuous beam of creep rupture. The steel spiral springs were periodically monitored to make the cross section bending moment within the given range of error.

In order to make a contrast analysis on the tensile properties of GFRP bars embedded in conrete beams with working cracks under different enviroments, specimens were subjected to three environmental conditions for 6, 12 and 18 months. The enviromental conditions were divided into three main types as follows : (1) specimens exposed to indoor natural environment as reference samples ; (2) accelerated aging of specimens aged in 50 ℃ alkaline solution with pH value of 12.7 ; (3) accelerated aging of specimens aged in 50 ℃ tap water to simulate high humidity environment of application. The test program is summarised in Table 2.

Table 2. Test program

Environment	Work cracks	Load level	Name	Temperature (℃)	Exposure time (month)		
					6	12	18
Controlled	Non pre-crack	0	RE	—	—	—	—
Alkaline solution	Pre-crack	25%	APS25	50	4	4	4
	Non pre-crack	0	AS0		4	4	4
Tap water	Pre-crack	25%	TPW25	50	4	4	4
	Non pre-crack	0	TW0		4	4	4

2.3 Test instruments

After the end of each accelerated exposed period, the beams were unloaded and the GFRP embedded bars in the tested beams were carefully extracted for tensile tests. The overall length of each bar was

1100 mm. To prevent shear failure on both ends of the GFRP bar, a 250 mm long steel pipe anchorage was fixed on the GFRP bar ends. The tensile tests were conducted in a SHT4106-G type microcomputer controlled electro-hydraulic servo universal testing machine according to the relevant requirements of ACI440.3R-04 B2. The applied load for each bar was recorded during the test with a data acquisition system monitored by a computer. In addition, scanning electron microscopy (SEM) and differential scanning calorimetry (DSC) were performed to observe and analyse the possible degradation of glass fibres and polymer matrix.

3 Results and discussion

3.1 Tensile property

The results of the tensile tests on the GFRP Bar showed an approximately linear behaviour with brittle failure. Specimen failed through the rupture of fibres, the failure was accompanied by delamination of fibres and resin. A similar failure mode was observed for GFRP bars embedded in concrete specimen.

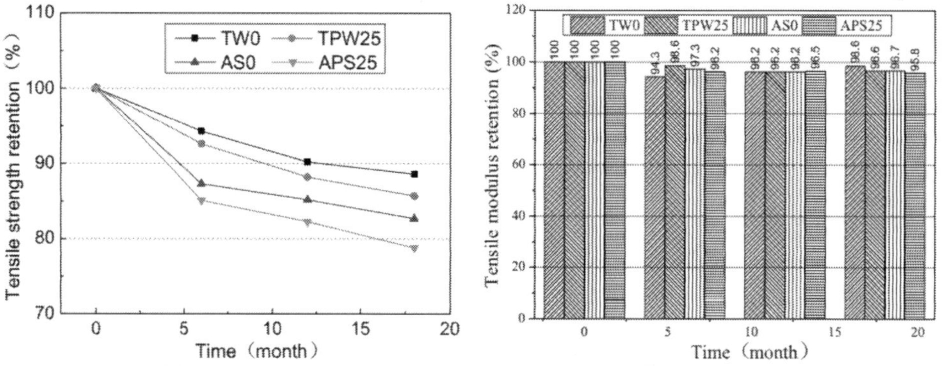

(a) tensile strength (b) tensile modulus

Figure 3. Change of tensile property of GFRP bars under different environments

In order to compare the tensile strength before and after accelerate aging test, the paper take the retention ratio as evaluation index. Figs. 3 show the tensile strength retention and tensile modulus retention, respectively. It can be seen from the Fig. 3(a) that the influence on the tensile strength of GFRP exposed in alkaline solution was significant greater than that of exposed in tap water. For the APS25 specimens, the tensile strength retention were decreased by 7.5%, 5.9% and 6.8% respectively after 6, 12 and 18 months of exposure than TPW25 specimen. It can be explained that the OH- accelerated the degradation rate of GFRP bar.

After 6, 12 and 18 months of exposure, the tensile modulus of AS0 specimen loss was 2.7%, 3.8% and 3.3% respectively. Compared with AS0 specimen, the modulus loss slightly increased to 1.1, 0.3% and 0.9%, respectively. For all environments, a slight decrease ranging between 0% and 6% was observed. These results are in good agreement with the experiments conducted by different researchers.

3.2 Microstructural analysis

DOI: 10.1051/mateccon f/20178802005

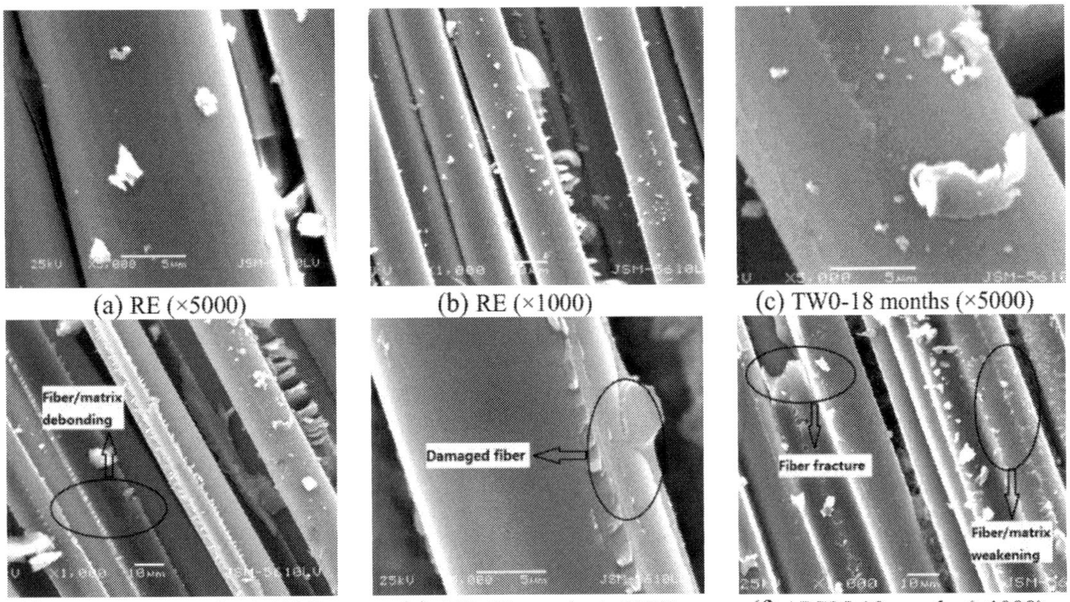

(a) RE (×5000) (b) RE (×1000) (c) TW0-18 months (×5000)

(d) TPW25-18 months (×1000) (e) AS0-18 months (×5000) (f) APS25-18 months (×1000)

Figure 4. Micrographs of the GFRP bars

The microstructure of GFRP bars before and after the aging tests can be clearly observed by SEM images. As shown in Fig.4 (a) and 4(b), there was almost no deterioration in the glass fibres, while the matrix around the fibre in both the TW0 and AS0 specimen were seriously affected by the environment. No damage occurred in the GFRP bars that embedded in concrete without work cracks and sustained load (Fig.4 (c) and 4(e)). However, with the increase of sustained load, part of the fibre fracture and the delamination at the interface of fibre and matrix in both TPW25 and APS25 (Fig.4 (d) and 4(f)) specimen. It is worth mentioning that debonding at the interface are the most common phenomena occurring in a GFRP material under sustained load in adverse environment.

3.4 Differential Scanning Calorimetry

Table 3. The glass transition temperature of GFRP bar

Conditioning	Temperature (℃)	Duration (m)	T_g run 1 (℃)	T_g run 2 (℃)	Cure ratio (%)
Controlled	—	—	124	126	98.4
TPW25	50	18	120	124	96.8
APS25	50	18	118	123	96.0

Table.3 presents the Glass transition temperature (T_g) for the first and second heating scans of the controlled TPW25, and APS25 samples. It is noticed that for all tested samples, the T_g values corresponding to the second heating run are higher than that of the first. This shift indicates that the samples had not been fully cured. The results in Table 3 also show no significant changes in T_g values for the specimens APS25. This indicates that DSC analysis detected no major impact effect on the resin's thermal properties as the result of the conditioning.

4 Conclusions

Based on the results of this study, the following conclusions can be drawn:

(1) When the specimen exposed in the same corrosive solution, the degradation rate of GFRP bar embedded in specimen with working cracks was faster than that of without cracks. While the work

cracks has no significant influence on the tensile elastic of the GFRP bar, the degradation rate of elastic modulus was no more than 5% in the whole test process.

(2) The micro-cracks in the concrete increased the effect of the external corrosion environment on the durability properties of the GFRP bar, but it cannot change the mechanisms of the mechanical degradation of the GFRP bar. No changes of the T_g occur as observed by differential scanning calorimetry.

(3) Compared with the deterioration of resin matrix, the deterioration of glass fibre and the delamination at the interface is more pronounced for GFRP bar embedded in concrete environments.

Acknowledgments

This work is financially supported by the National Nature Science Foundation of China with Grant No. 51178361. The authors wish to gratefully thank the Material Research and Testing Center of Wuhan University of Technology for their assistance.

References

1. A. Yousef., El-gamal Sheriff, H.A Tarek, et al. Effect of harsh environmental conditions on the tensile properties of GFRP bars. Composite. Part B: Engineering, (2013), 45(1): 835-844.
2. A. Mufti, N. Banthia, B. Benmokrane, et al. Durability of GFRP composite rods. Concr Inter, (2007), 29(2):37-42.
3. F. Micelli, A. Nanni. Durability of FRP rods for Concrete Structures. Constr Build Mater. (2004), 18(7):491-503.
4. A. Mukherjee, S.J Arwikar, Performance of glass fiber reinforced polymer reinforcing bars in tropical environments part I: Structural scale tests. ACI Structural J, (2005), 102(5): 745-753.
5. Y. Chen, J.F. Davalos, I. Ray, and H Y Kim. Accelerated aging tests for evaluations of durability performance of FRP reinforcing bars for concrete structures. Compos Struct. (2007), 78(1), 101-111.
6. C.C Li, A.M Yu, Y.L Wang. Durability of FRP rebars in alkaline environment of concrete. Build Sci, (2013), 29(1), 47-51. (in Chinese)
7. W. Wang, W.C Xue. Accelerated aging tests for evaluations of tensile properties of GFRP bars exposed to alkaline solution. J Build Mater, (2012), 15(6):760-766.
8. M. Robert, P. Cousin, B. Benmokrane. Durability of GFRP reinforcing bars embedded in moist concrete. J Compos Constr, (2009), 13(2): 66-73.
9. C. Sun, W.H Qiu, X.B Zeng. Experimental research on long-term mechanical behavior of GFRP bars in different environmental conditions. FRP/CM. (2014), 8:88-91.
10. A. Debaiky, G. Nkurunziza, B. Benmokrane, et al. Residual tensile properties of GFRP reinforcing bars after loading in severe environments. J Compos Constr, (2006), 10:370-380.
11. M. Robert, B. Benmokrane. Combined Effects of Saline Solution and Moist Concrete on Long-term Durability of GFRP Reinforcing Bars. Constr Build Mater, (2013), 38, 274-284.
12. G.Wu, Y. Zhu, Z.Q Dong et al. Experimental study on the corrosion resistance performance of BFRP bars in the alkaline environment. China Civil Engine J. (2014), 47(8):32-41.
13. L. Dai, X.J He, W.R Yang. Experimental study on the durability of GFRP bars embedded in actual concrete environment. FRP/CM. (2015), 8:75-79.

MATEC Web of Conferences **88**, 02006 (2017) DOI: 10.1051/matecconf/20178802006
CBNCM 2016

Transition Curvature of Soft Matrix Composite Laminates during Bending

Z.Y. Xiong[1,a], R. X. Song[1], R. Wu[1], Z.X. Kang[1], Q. Liu[1] and H.J. Zhang[1]

[1]*Beijing Municipal Institute of Labour Protection, Beijing 100054, China*

Abstract. As an Elastic memory composite (EMC) laminate is bent, the prebuckled plate theory is valid until the transition curvature at which point the postbuckled solution becomes effective. This is the point at which microbuckling is predicted to occur. From the point onward, EMC laminate follows the postbuckled solution. In this paper, a solution of the transition curvature is proposed by combining matrix shear energy with Timoshenko's elastic foundation energy. By comparing with the intersection point of the straight line given by the prebuckled plate theory and the curve given by the postbuckled solution, the solution can be used to predict the point at which microbuckling is predicted to occur.

1 Introduction

Elastic memory composites (EMCs) are similar to the conventional fiber-reinforced composites except matrix material having a shape-memory function as compared with the conventional polymer matrix. Due to some valuable properties, such as high folding deformation, very high deployed stiffness- and strength-to-weight ratios, low density[1], EMCs have attracted considerable interests in the applications of deployable space-structure industry. Several research teams have reported successful fabrications of various components for deployable structures using EMCs, such as laminated plates and shells, pultruded rods, open-grid lattices, and hinges[2-4].

Compared with the ultimate strain of the reinforcing fibers, the folding strain of continuous-fiber-reinforced EMCs is much higher. The main mechanism that allows much higher folding strain without damage is due to microbuckling, and post-microbuckling response of the compressively loaded fibers during bending, which have been confirmed by tests[5-7]. For example, although the ultimate strain of high-performance carbon fibers is usually lower than 1%, continuous-carbon-fiber-reinforced EMC laminate can receive more than 5% equivalent bending strain.

Fig. 1 is the sketch of microbuckling in EMC laminate during bending. Microbuckling is regarded as an elastic instability response of the compressively loaded reinforcement fibers. For EMCs under bending, as the temperature is higher than the glass transition temperature of the matrix (T_g), the matrix will become compliant. The compliant matrix can not supply enough shear stiffness to prevent the fibers from microbulking under the pressure. Therefore the reinforced fibers in the compressive zone can achieve much higher deformation without damage. In the meantime, microbuckles will dramatically decrease the stiffness of the fibers in compressive zone and induce their values to be much lower than that in tensile zone. Since the matrix is compliant and can not endure much higher

[a] Corresponding author : 13671108368@126.com

load, the mismatch of the stiffness between the compressive and tensile zones will cause the neutral axis of EMC laminate significantly shifting toward the tensile zone. By such deformation adjustment, the maximum tensile strain of the reinforced fibers can also keep under an endurable value even if the equivalent bending strain of EMC laminate is very high.

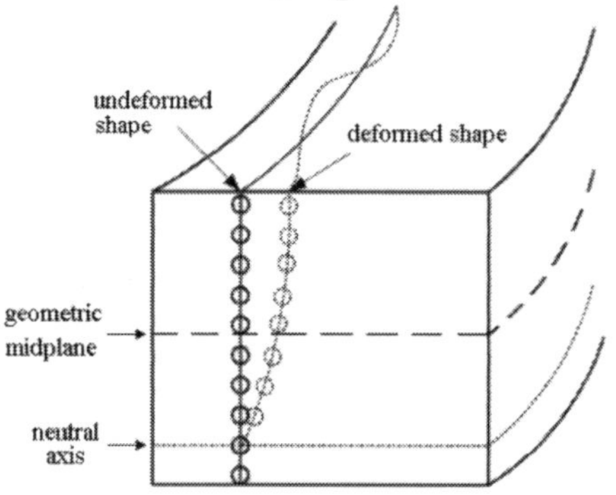

Figure 1. Sketch of microbuckling in EMC laminate during bending.

As an EMC laminate is bent, the prebuckled plate theory can be expressed as

$$M_{linear} = \frac{E_x t^3 w\kappa}{12\left(1 - v_{xy} v_{yx}\right)} \tag{1}$$

where w is the plate width, κ is the bending curvature, E_x is the longitudinal composite Young's modulus, t is the thickness of the laminate, v_{xy} and v_{yx} are respectively the axial, transverse in-plane composite Poisson's ratio. the postbuckled solution given by William H. Francis IV[8] is expressed as the following

$$M = \frac{G_{xy} t^2 w}{12\varepsilon_b^2}\left[\left(\frac{G_{xy}}{E_x} - 2\varepsilon_b\right)\sqrt{\left(\frac{G_{xy}}{E_x}\right)^2 + 4\frac{G_{xy}}{E_x}\varepsilon_b} - \left(\frac{G_{xy}}{E_x}\right)^2 + 6\varepsilon_b^2\right] \tag{2}$$

where G_{xy} is the in-plane composite shear modulus, ε_b is the equivalent bending strain defined by

$$\varepsilon_b = t\kappa/2 \tag{3}$$

ε_b is not the actual strain on either the compression or tension surface because the calculation assumes the neutral strain surface is located at the geometric midplane. However, it is a reasonable indication of the magnitude of the deformation created by the bending. In order to express the moment in terms of the plate curvature rather than the equivalent bending strain, substituting Equation (3) into Equation (2) gives

$$M = \frac{G_{xy} w}{3\kappa^2}\left[\left(\frac{G_{xy}}{E_{xx}} - t\kappa\right)\sqrt{\left(\frac{G_{xy}}{E_{xx}}\right)^2 + 2t\kappa\frac{G_{xy}}{E_{xx}}} - \left(\frac{G_{xy}}{E_{xx}}\right)^2 + \frac{3}{2}(t\kappa)^2\right] \tag{4}$$

The prebuckled plate theory, Equation (1), is valid until the transition curvature at which point the postbuckled solution, Equations (2) or (4), becomes effective. Therefore, this is the point at which microbuckling is predicted to occur. From the point onward, EMC laminate follows the postbuckled solution. The present article aims to provide an analytical solution to predict the transition curvature of EMC laminate under bending. The solution and the intersection point of the straight line given by Equation (1) and the curve given by Equation (4) are compared, and the reasons of errors are also reasonably analyzed.

2 Analytical solution of transition curvature

2.1 Critical microbuckling strain

By combining the matrix strain energy given by Rosen's solution for shear instability-mode microbuckling composite[9] with Timoshenko's elastic foundation energy[10], a suitable model and solution is proposed to determine the critical wavelength and critical microbuckling strain of the fiber-microbucking in EMC laminate under pure bending. Based on the laboratory observations that the same wavelength of sinusoidal-like and in-plane microbuckling exist in the EMC laminate under pure bending[6-7], the following simplified assumptions are applied: all fibers display sinusoidal-like microbuckles with the same mode number; the microbuckling mode is recognized as a 1D mode instead of 3D mode since no out-of-plane microbuckle is observed in tests; only shear microbuckling is identified as the available tests exhibit that it is the dominated mode for EMC laminates under bending.

Fig. 2 shows the 1D microbuckling mode used in calculating matrix shear energy

$$U_m = \int_V \frac{1}{2} G_m \gamma_{xy}^2 dV \tag{5}$$

where γ_{xy} is the shear strain, and can be determined from Fig. 2 using the following expression

$$\gamma_{xy} = \frac{dy}{dx} + \frac{h}{2c}\frac{dy}{dx} \tag{6}$$

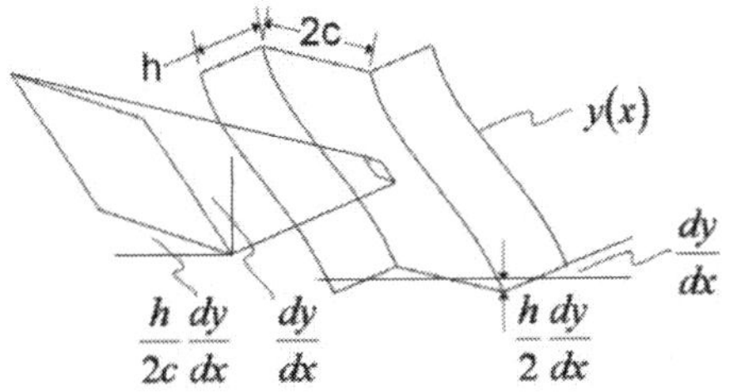

Figure 2. 1D representation of post-microbuckled EMC laminate.

Since a fiber in EMC laminate under bending shows a sinusoidal-like microbuckling shape, it can be described as

$$y = a\sin(m\pi x/l) \tag{7}$$

where m is the mode number, namely, number of waves along the fiber length and must be a positive integer to assure the boundary condition, a is the amplitude, and l is the length of the laminate.

Substituting Equation (7) into Equation (6), then Equation (6) into Equation (5) and the integrating result gives the elastic foundation strain energy for specimen with thickness h

$$U_m = \frac{G_m hc\pi^2}{2l}\left(1+\frac{h}{2c}\right)^2 a^2 m^2 \tag{8}$$

where G_m is shear modulus of the matrix, c is half-spacing between adjacent fibers, and h is fiber diameter.

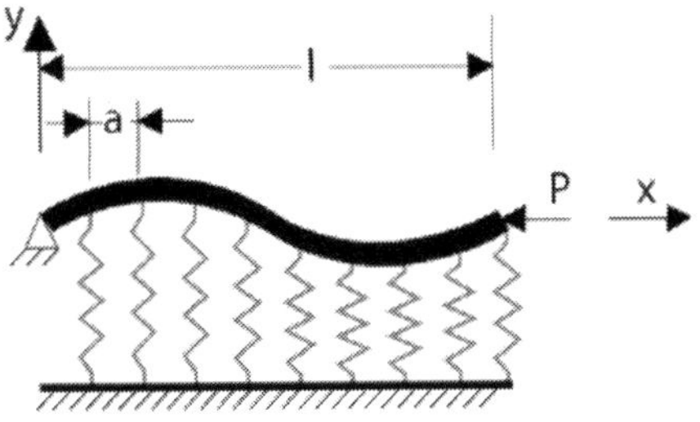

Figure 3. Buckling of a compressively loaded bar on an elastic foundation.

Fig. 3 shows Timoshenko's solution for buckling of a compressively loaded bar in an elastic foundation. The elastic foundation is modeled by a number of springs with a stiffness of α separated by a distance d. The stiffness per unit length of the elastic foundation can be expressed as

$$\beta = \alpha / d \tag{9}$$

Therefore the strain energy stored in the elastic foundation is expressed as

$$U_s = \int_0^l \frac{1}{2} \beta y^2 dx = \beta l a^2 / 4 \tag{10}$$

The stiffness β of the elastic foundation in a unit length is determined by combining Equations (10) and (8)

$$\frac{G_m h c \pi^2}{2l} \left(1 + \frac{h}{2c}\right)^2 a^2 m^2 = \frac{1}{4} \beta l a^2 \tag{11}$$

from which the following expression can be obtained

$$\beta = \frac{2hc\pi^2 G_m}{l^2} \left(1 + \frac{h}{2c}\right)^2 m^2 \tag{12}$$

The solution for the critical load at which buckling occurs in the mode of Fig. 3 is given by Timoshenko[10] as

$$P_{cr} = \frac{\pi^2 E_f I_f}{l^2} \left[m^2 + \frac{\beta l^4}{m^2 \pi^4 E_f I_f} \right] \tag{13}$$

where E_f is fiber elastic modulus, and I_f is fiber area moment of inertia.

The minimized P_{cr} is obtained by differentiating Equation (13) with respect to m and the result is

$$\beta = \frac{\pi^4 m^4 E_f I_f}{l^4} \tag{14}$$

The Equations (14) and (12) yield a new Equation (15):

$$\frac{2hc\pi^2 G_m}{l^2} \left(1 + \frac{h}{2c}\right)^2 m^2 = \frac{\pi^4 m^4 E_f I_f}{l^4} \tag{15}$$

Hence the microbuckling wavelength can be obtained as

$$\lambda = \frac{l}{m} = \sqrt{\frac{\pi^2 V_m E_f h^3}{24 c G_m}} \tag{16}$$

where V_m is matrix volume fraction, and it is equal to 2c/(2c+h).

Substituting Equations (14) and (16) into Equation (13), the critical load P_{cr} can be determined as

$$P_{cr} = \frac{4chG_m}{V_m^2} \tag{17}$$

Therefore the critical stress and strain are respectively obtained as

$$\sigma_{cr} = \frac{4chG_m / V_m^2}{h(h+2c)} = \frac{2G_m}{V_m} \tag{18}$$

$$\varepsilon_{cr} = \frac{2G_m}{V_m E_f} \tag{19}$$

Equation (16) gives a solution of microbuckling wavelength. The critical compressive stress and strain on the initiation of the microbuckle are respectively determined by Equations (18) and (19).

The traditional microbuckling solutions have two fundamental buckling modes, namely shear buckling[9] and extension-buckling[11] modes. The theoretical buckled shape for shear mode is a single half sine function along the length of the specimen of EMC laminate under bending. However, the test specimen is 250 mm in length and the fibers have a normal wavelength of around 1 mm[7]. In addition, the critical microbuckling wavelength predicted by the extension mode is about 0.09 mm. Hence, both shear- and extension-mode solutions in the classical fiber microbuckling mechanics are ineffective in forecasting the bending deformation of EMC laminates. Campbell et al.[12] provides a theoretical calculation of microbuckling wavelength in EMC laminates. The wavelength value is

1.035 mm, which is in good agreement with the test result, 1 mm. However, according to the expression of microbuckling wavelength the derived critical microbuckling stress goes down with increasing the fiber volume fraction, which is difficult to comprehend and questionable. By assuming a linear relation between the fiber microbuckling amplitude and the thickness of the laminate, Campbell and Maji[13]further proposed an improved 2D microbuckling model. Unfortunately, the similar shortage still exists. Francis et al.[14,8] proposes a model describing the complex post-microbuckled behavior for EMC laminate under bending. However, this study focuses on post-microbuckled behavior and the model does not apply for small deformations.

Substituting those test parameters[7] into Equation (16) deduced in this study, the microbuckling wavelength of the continuous-carbon-fiber/resin EMC laminate is about 0.72 mm, which is comparable with the test result, 1 mm. Meanwhile, the expressions of the critical microbuckling stress, Equation (18), and strain, Equation (19), derived from the expression of microbuckling wavelength, Equation (16), in the proposed model are reasonable and intuitively correct.

2.2 Solution of transition curvature

As an EMC laminate is bent, it is assumed that it's behavior obeys the lowest energy solution path. the prebuckled laminate theory is valid until the transition curvature at which point the postbuckled solution becomes the lower energy solution. Equation (1) is only applicable to linear laminate and is, thus, referred to as the prebuckled solution to differentiate it from the postbuckled solution provided in Equation (4). Therefore, this is the point at which microbuckling is predicted to occur. From the bending curvature onward, the EMC laminate follows the postbuckled solution.

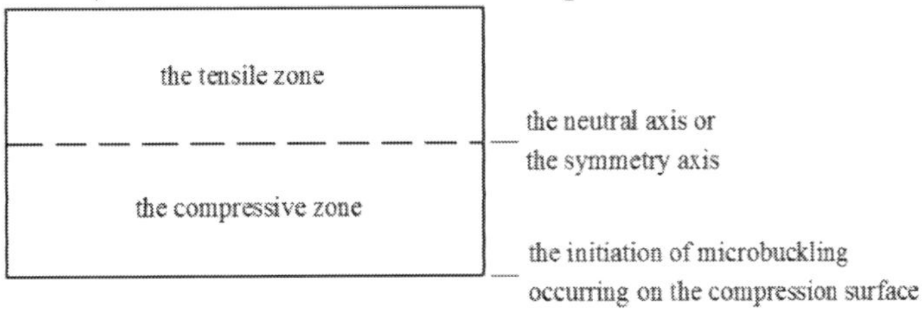

Figure. 4 Sketch of cross section of EMC laminate under transition curvature

Fig. 4 is the sketch of cross section of EMC laminate under the transition curvature. Soft Matrix composite laminate behaves equivalently in tension and compression implying that the neutral axis will reside at the symmetry axis of cross section of laminate. The initiation of microbuckling occurs on the compression surface of the EMC laminate and the other parts of the EMC laminate have not microbuckled. The strain on the compression surface of the EMC laminate can be expressed as

$$\varepsilon = \frac{t\kappa_{tr}}{2} \tag{20}$$

where κ_{tr} is the the transition curvature.

Equations (20) and (19) induce a new Equation (21)

$$\frac{t\kappa_{tr}}{2} = \frac{2G_m}{V_m E_f} \tag{21}$$

Hence, the transition curvature of soft matrix composite laminate during bending will be obtained as

$$\kappa_{tr} = \frac{4G_m}{tV_m E_f} \tag{22}$$

The transition curvature, κ_{tr}, is proportional to the matrix shear module, G_m. κ_{tr} increases with the increase of G_m and decreases with the decrease of G_m. κ_{tr} is inversely proportional to thickness of compoite t, matrix volume fraction V_m, and fiber Young's modulus E_f. Those are understandable and intuitively correct.

3. Verification of solution

Assigning the parameters in Equations (1) and (4) the values given in Table 1 and plotting the moment applied across the laminate length of interest as a function of the bending curvature produces the plot shown in Figure 5. The straight line is given by Equation (1) and the curve is given by Equation (4) in Figure 5. However, due to the absence of a reliable in-plane shear modulus of composite G_{xy}, G_{xy} is treated as a curve fitting parameter and it is found that a value of G_{xy} which is lower than the value in Table 1 by a factor of 1.9 is needed for data/model correlation[8].

Table 1. Related parameters of soft matrix laminate[8]

Symbol	Unit	Value	Description
E_x	Pa	106×10^9	longitudinal composite Young's modulus
G_{xy}	Pa	67.4×10^6	in-plane shear modulus of composite
v_{xy}	-	0.264	axial in-plane composite Poisson's ratio
v_{yx}	-	0	transverse in-plane composite Poisson's ratio
t	mm	0.8, 1.5, 3	thickness of compoite
w	mm	25	Width of compoite
G_m	psi	3000	matrix shear module
V_m	-	0.60	matrix volume fraction
E_f	Pa	276×10^9	fiber Young's modulus

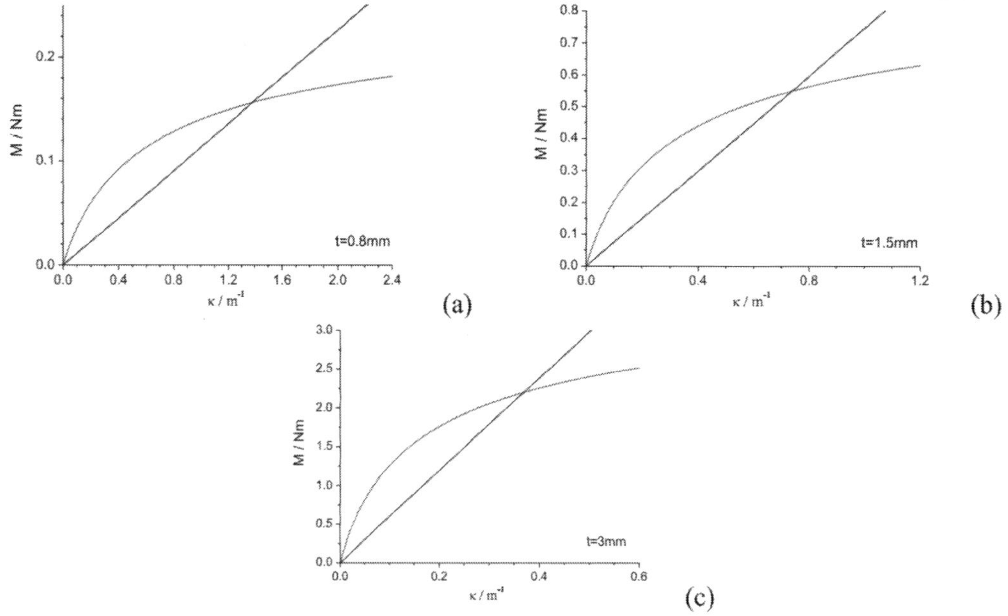

Figure 5. The intersection point of the straight line and the curve (a) t=0.8mm,(b) t=1.5mm, (c) t=3mm

The strain energy stored within the EMC laminate as it is bent is proportional to the area under the curves in Figure 5. Therefore, the lowest energy solution at a given curvature is the solution appearing lowest in the vertical direction. As an EMC laminate is bent, it is assumed that it's behavior follows the lowest energy solution path. Equation (1) is valid until the transition curvature at which point the postbuckled curve becomes the lower energy curve. Therefore, this is the point at which microbuckling is predicted to occur. From the transition curvature onward, the laminate follows the postbuckled solution, Equation (4). From Figure 5, we can know that the intersection points of the straight lines and the curves are respectively about 1.37, 0.73, 0.36 m^{-1} under three different thicknesses of laminate (0.8, 1.5, 3.0mm).

In keeping with the requirement of Reference[8], the experiment values will be used to represent the material properties rather than the predicted values and the experiment values usually have higher reliability. The value of matrix shear module G_m in Table 1 is not the predicted value given by Reference[8] rather than the value given by Reference[15]. Substituting the relevant data in Table 1 into Equation (22), the results are approximately 0.625, 0.333, 0.167 m^{-1} respectively under three different thicknesses of laminate (0.8, 1.5, 3.0mm).

The transition curvatures in Figure 5 are approximately twice the predicted values given by Equation (22). The possible reasons exist in two aspects: (a) devices for measuring hard matrix are now used to measure soft matrix, so the errors are certainly considerable; (b) a literature[6] shows that, the errors may be attributed to the apparent extreme temperature sensitivity of the matrix shear module G_m. A temperature of 95°C would result in a G_m that is approximately two and a half times the G_m corresponding to 100°C.

4. Conclusions

This paper analyses the transition curvature of soft matrix composite laminates during Bending. By combining Timoshenko's elastic foundation energy with the matrix strain energy given by Rosen's solution for shear-instability-mode microbuckling composite, a model is proposed to describe the microbuckling deformation of a fiber in soft matrix and the critical compressive strain on the initiation of the microbuckle is determined. In the meanwhile, compared with other models, the proposed model is more closed to the real situation. When the strain on the compression surface of the EMC laminate is equal to the critical compressive strain on the initiation of the microbuckle, the analytical solution of the transition curvature is derived. By comparing with the intersection points of the straight lines given by the prebuckled plate theory and the curves given by the postbuckled solution, the solution can be used to predict the point at which microbuckling is predicted to occur, and the causes of the errors are also analyzed in this paper.

Acknowledgement

This work was funded by Municipal Finance Project of Beijing Municipal Science and Technology Research Institute (PXM2015-178304-000021-008).

References

1. Lake, M.S., Munshi, N.A. and Tupper, M.L. Application of Elastic Memory Composite Materials to Deployable Space Structures[J]. AIAA Paper No.4602 (2001).
2. Gall, K., Mikulas, M., Munshi, N.A., Beavers, F. and Tupper, M. Development and Testing of a Hinge/Actuator Incorporating Elastic Memory Composites[J]. AIAA Paper No. 1496(2003).
3. Mark, S.L. and Hazelton, C.S. Development of Coilable Longerons Using Elastic Memory Composite Material[J]. AIAA Paper No. 1453 (2002).
4. Francis, W.H., Lake, M.S. and Mallick, K. Development and Testing of a Hinge/Actuator Incorporating Elastic Memory Composites[J]. AIAA Paper No. 1496(2003).
5. Call, K. Micro-mechanisms of Deformation in Fiber Reinforced Polymer Matrix Elastic Memory Composites[J]. AIAA Paper No. 1419(2001).
6. Campbell, D. and Maji, A.K. Deployment precision and Mechanics of Elastic Memory Composites[J]. AIAA Paper No. 1495(2003).
7. Francis, W.H. and Lake, M.S. A Review of Classical Fiber Microbuckling Analytical Solutions for Use with Elastic Memory Composites[J]. AIAA Paper No. 1764(2006).
8. William H. Francis IV. The Mechanics of Post-Microbuckled Compliant-Matrix Composites[D]. Master Thesis. Colorado: University of Colorado, (2008).
9. Rosen, B.W. Fiber Composite Materials[M], American Society for Metals, Metals Park, Ohio, Ch. 3.(1965)
10. Timoshenko, S., Theory of Elastic Stability[M], McGraw-Hill Book Co., Inc., pp109-112.(1936)
11. Dow, N. F., and Rosen, B. W. Evaluations of Filament-Reinforced Composites for Aerospace Applications[J] NASA CR-207(1965).

12. Campbell, D., Lake, M.S. and Mallick, K. A Study of the Compression Mechanics of Soft-resin Composites[C] In: The 45th AIAA/ASME/ASCE/AHS/ASC Structures, Structural Dynamics and Materials Conference, Palm Springs, California.(2009)
13. Campbell, D. and Maji, A.K. Failure Mechanisms and Deployment Accuracy of Elastic Memory Composites[J]. Journal of Aerospace Engineering, 19(3):184-193.(2006)
14. Francis, W.H., Campbell, D., Lake, M., Dunn, M. and Qi, H.J. Elastic Memory Composite Microbuckling Mechanics: Closed-Form Model with Empirical Correlation[J]. AIAA Paper No.2164(2007).
15. Douglas Campbell, Mark S Lake, Kaushik Mallick. A study of the bending mechanics of elastic memory composites [C]. 45th AIAA/ASME/ASCE/AHS/ASC Structures, Structural Dynamics & Materials Conference19-22 April 2004, Palm Springs, California.(2009)

Gold Magnetic Nanoparticles-based Chemiluminescent Immunoassay for Detection of Chloramphenicol in Milk

Linyu Wang [1], Manwen Yao [2], Xiangyi Fang[3], Peng Tang[3] and Xi Yao[1,a]

[1] *School of Electronic and Information Engineering, Xi'an Jiaotong University, Xi'an, PR China*
[2] *Tongji University, Shanghai, PR China*
[3] *School of Science, Xi'an Jiaotong University, Xi'an, PR China*

Abstract. chemiluminescent immunoassays (CLIA) based on gold magnetic nanoparticles (Au-MNPs) were developed for rapid analysis of chloramphen icol (CAP) in milk sample. Anti-CAP antibodies were immobilized on the surfaces of Au-MNPs, luminol (Method I) and 2',6'-DiMethylcarbonylphenyl-10-sulfopropylacridinium-9-carboxylate 4'-NHS Ester, (NSP-DMAE-NHS, Method II) were exploited in competitive CLIA for CAP detection in milk using a homemade luminescent measurement system. The sensitivities and limits of detection (LODs) of the two methods were obtained according to the inhibition curves. It indicated that NSP-DMAE-NHS as luminescence reagent (reaction II) was more sensitive and effective than luminol (reaction I). The LOD of reaction II reached 0.008 ng/ml while it was 4 ng/ml in reaction I. Moreover, the linear range of the inhibition curve of the former was wider than that of the latter. Such results indicated that the proposed CLIA stratery employing NSP-DMAE-NHS was more sensitive than other immunoassay method for CAP detection.

1 Introduction

Chloramphenicol (CAP), a kind of cheap and effective bacteriostatic broad-spectrum antibiotics, was widely used in animal husbandry to treat kinds of sensitive bacteria infection. However, it was found that CAP had serious adverse effects against human bodies, leading to bone marrow depression and even aplastic anemia [1-2]. Due to these side effects, it was banned from use in animal feeds and food in the European Union (EU), Switzerland, USA and other countries [3-4].

Various methods have been developed for detection of CAP residues in animal food. Conventional methods like microbiological methods [5] were not sensitive for CAP and were totally obsolete. Gas/liquid chromatography (GC/LC) [6-7] and gas/liquid chromatography-mass spectrometry (GC-MS/LC-MS) [8-10] are internationally recognized but expensive, complicated to operate, and not suitable for detecting lots of samples. Immunoassays including radio immunoassay (RIA) [11], enzyme-linked immunosorbent assay (ELISA) [12-13] have been reported frequently to screen CAP. RIAs have high sensitivity but exists radioactive waste, and ELISAs have shortcoming of higher limit of detection (LOD). Chemiluminescent immunoassays (CLIAs) have some features such as non-radioactive wastes, simple instrumentation, the very low limit of detection (LOD), high sensitivity, wide dynamic range and rapid analysis time in only several seconds. Those characteristics make it broadly used in clinical, environment, food and factory analysis in recent years [14-15].

[a] Corresponding author : xyao@mail.xjtu.edu.cn

Magnetic nanoparticles had been widely used as solid carry in RIA, ELISA and CLIA, due to: (i) the easier manipulation because the magnetic nanoparticles can be controlled in magnetic field instead of centrifuging repeatedly, (ii) the increased surface area as well as (iii) the faster assay kinetics because the nanoparticles are in suspension [16]. In addition, due to the improved washing and separation steps, the matrix effect is minimized in complex samples [17]. Gold magnetic nanoparticles (Au-MNPs) have outstanding characteristics with combining the superparamagnetism of magnetic nanoparticles and the ability of being modified by biomolecules on gold surface.

Chemiluminescence reactions which exploited luminol (or N-(4-Aminobutyl)-N-ethylisoluminol, ABEI) and an enhancer (such as p-iodophenol) by hydrogen peroxide (H_2O_2) in the presence of horseradish peroxidase (HRP) were reported widely [18]. Weeks [19] synthesized acridinium esters (AE)-NHS [4-(2-Succinimidyfoxycarbonylethyl) phenyl-10methylacridinium -9-carboxylate fluorosulfonate] and it has high luminous efficiency with no need of catalyst and enhancer. However, the thermal stability of AE-NHS is not so good that many researches have been done to syntheses more stable acridinium esters (2',6'-DiMethylcarbonylphenyl-10-sulfopropylacridinium-9-carboxylate 4'-NHS Ester). Which is an ideal chemical luminescence immunoassay marker for the thermal stability and hydrolytic stability. This paper explored the Au-MNPs-based CLIA methods for CAP detection in milk samples respectively with luminol (reaction I) and NSP-DMAE-NHS (reaction II) as luminescence reagent.

2 Materials and methods

2.1 Synthesis of Fe_3O_4@SiO_2@Au magnetite nanoparticles

Fe_3O_4 Magnetic nanoparticles (Fe-MNPs) were prepared by solvothermal synthesis reaction with some modifications. Typically, $FeCl_3$ $6H_2O$ (1.35 g) was first dissolved in 40 mL ethylene glycol to form a clear solution, followed by addition of polyethylene glycol (1.0 g), afterward, sodium acetate (3.6 g) was added with stirring. The mixture sealed in autoclave. It was heated to 200℃ for 10 h. The black products were washed and dried in a vacuum oven at 80℃.

Fe-MNPs (340 mg) were dispersed in solvent that contained 250mL ethanol and 30 mL deionized water under ultrasonic vibration for five minutes. Then 3 mL ammonia (28 wt.%) and 2 mL TEOS were added slowly with strring and the reaction was continuous stirred for another 15 hours. We separated the Fe_3O_4@SiO_2 with a magnet. Then, we used APTES as a amination reagent. Fe_3O_4@ SiO_2 (150mg) in suspension under ultrasonic vibration in 50ml ethanol were added 100 uL APTES, which reacted 48 hours at room temperature. The solid products were collected, washed with deionized water and ethanol and re-dispersed in 100 ml water.

Gold colloids with the size of 30 nm were prepared by reduction of $HAuCl_4$ with sodium citrate. Then, 30 mL of the amino-functionalized Fe_3O_4@SiO_2 nanoparticles dispersion was added dropwise to 100 ml of the citrate-stabilized gold colloids. With stirring for 1 hour, the resulting product was separated magnetically and washed with water.

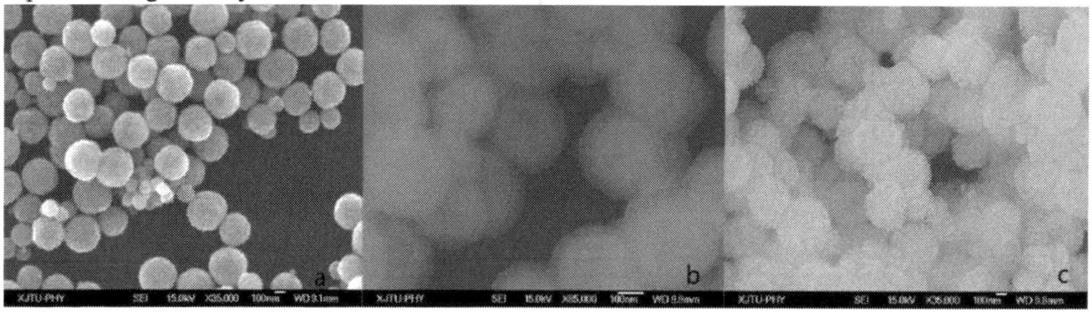

Figure.1. a SEM of Fe_3O_4 b SEM of Fe_3O_4@SiO_2 c SEM of Fe_3O_4@SiO_2@Au

2.2 Anti-CAP antibody immobilization on Au-MNPs

The immobilization process of the anti-CAP antibodies on the Au-MNPs is shown in Fig 3(A). Carboxyl-(Au-MNPs) which was formed by immersing Au-MNPs in a solution of 11-Mercaptoundecanoic acid (MUA, 5 mM) in ethanol for 16h at room temperature was washed twice with MES (2-(N-Morpholino) ethanesulfonic acid) buffer (25 mM, pH5.12). Afterwards, the carboxyl-(Au-MNPs) was re-suspended in MES buffer and activated by incubation with a 1:1 ratio mixture of 200 mM EDC (N-ethyl-N'-(3-dimethylaminopropyl) carbodiimide hydrochloride) and 50 mM NHS (N-hydroxysulfosuccin-imide) in MES buffer for 30 min. After washed three times with MES buffer and distilled water respectively, 100 μl of the Au-MNPs were incubated with 100 μl of anti-CAP antibody at 37 °C for 2 hours. After incubation, the supernatant was removed and washed three times with PBS (0.01M, pH7.4) in a magnetic field.

2.3 Determination of the tracers' optimized amount

Before construction of standard curves for different concentration of CAP in two chemiluminescence reactions, the optimized amount of tracers (CAP-HRP and CAP-(rabbit IgG)-(NSP-DMAE-NHS) conjugates) could be determined by binding them on the antibody-immobilized Au-MNPs with different concentrations. Fig.2 (A) shows the emission intensity (CPS) of the chemiluminescence corresponding to different amount of CAP-HRP. The concentration of CAP-HRP was 0.6 mg/ml. It can be seen that the chemiluminescence intensity increased with the increase of CAP-HRP amount in the range of 50-200 μl and the CPS reached saturation at 200 μl. This extraordinary result may be explained by steric effect. In this case, CAP-HRP saturation amount was 200 μl. In consideration of ensuring higher test sensitivity in the competitive reaction and tracer consumption, the optimal CAP-HRP amount of 180 μl was selected as the fixed volume for competitive reaction.

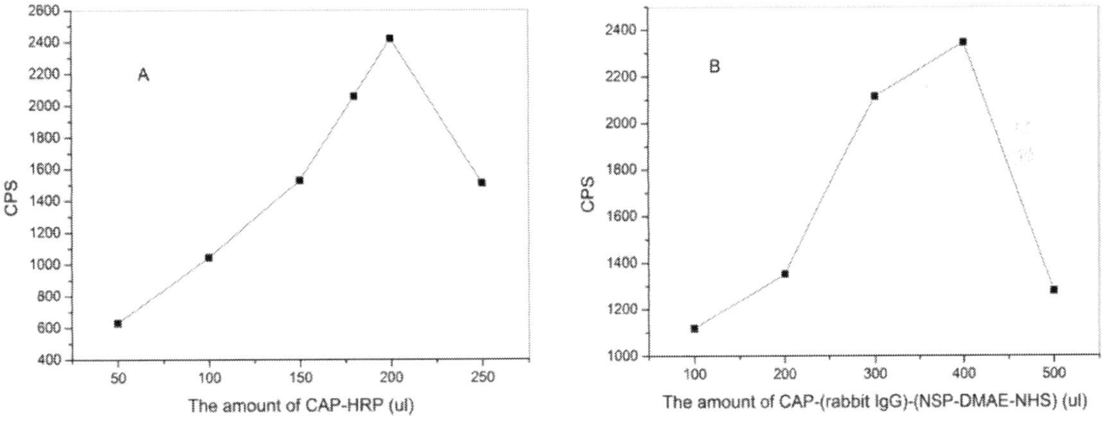

Figure.2. (A) The emission intensity versus the amount of CAP-HRP (B) The emission intensity versus the amount of CAP-(rabbit IgG)-(NSP-DMAE-NHS)

Fig.2 (B) shows the CPS according to different amount of CAP-(rabbit IgG)-(NSP-DMAE-NHS). There also existed the steric effect and the saturation amount of this tracer was 400 μl. In consideration of the same reason, the optimal amount of CAP-(rabbit IgG)-(NSP-DMAE-NHS) of 300 μl was determined for competitive reaction in target CAP measurement.

2.4 Establishment of CLIA method based on Au-MNPs for CAP detection

The process of reaction I is shown in Fig.3 (B). CAP succinate sodium and commercial milk powder were dissolved in deionized water to prepare the standard milk samples in which the concentration of the milk powder was 1mg/ml and the concentrations of the CAP succinate sodium were 0.01, 0.1, 1, 10, 100 and 1000 ng/ml respectively. The immobilized anti-CAP antibodies on Au-MNPs, which were prepared by the procedures described in the previous section, were allowed to react with a mixture of certain amount of CAP-HRP tracer and the spiked milk samples. The mixture was allowed to incubate at 37 °C for 2 hours with gentle shaking. Then the Au-MNPs were washed with PBS in magnetic field. In this study, the substrate solution was made of luminol (0.5 mM), p-iodophenol (0.7 mM) and H_2O_2 (14 mM) in Tris-HCl buffer (20 mM, pH8.6) according to a former experiment. A volume of 100 μl of the mixture of luminol and p-iodophenol (1:1) was added to the Au-MNPs, followed by the injection of 100 μl of H_2O_2 by a peristaltic pump to induce chemiluminescence of the antibody-bound CAP-HRP on the Au-MNPs. At the same time, the chemiluminescence emission was measured by the luminescent measurement system.

The process of reaction II is shown in Fig.3 (C). The preparation of spiked milk samples was the same as reaction I except that the concentrations of the CAP succinate sodium were 0.0001, 0.001, 0.01, 0.1, 1, 10 ng/ml respectively. CAP-(rabbit IgG)-(NSP-DMAE-NHS) conjugates immobilized on Au-MNPs. Then the particles were washed by PBS in a magnetic field. Here, HNO_3+ H_2O_2 and NaOH acted as the pre-trigger and trigger respectively to induce chemiluminescence emission. A volume of 100 μl of the pre-trigger was previously added to the Au-MNPs and 100 μl of trigger was injected into the tube by a peristaltic pump to induce chemiluminescence of the antibody-bound conjugates of CAP-(rabbit IgG)-(NSP-DMAE-NHS) on Au-MNPs. At the same time, the chemiluminescence emission intensity was measured versus time by the luminescent measurement system.

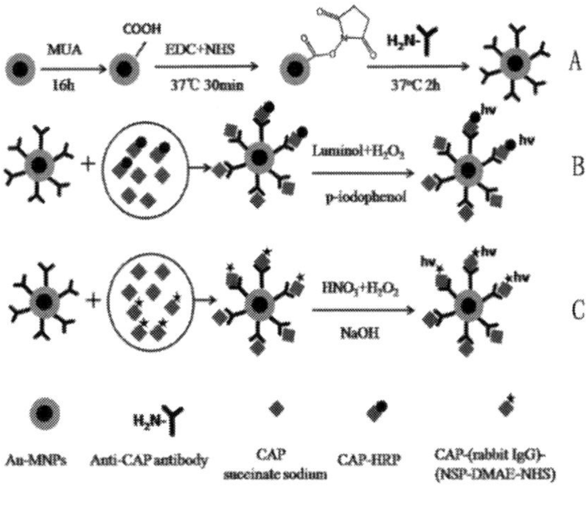

Figure.3. two CLIA reactions for CAP detection.

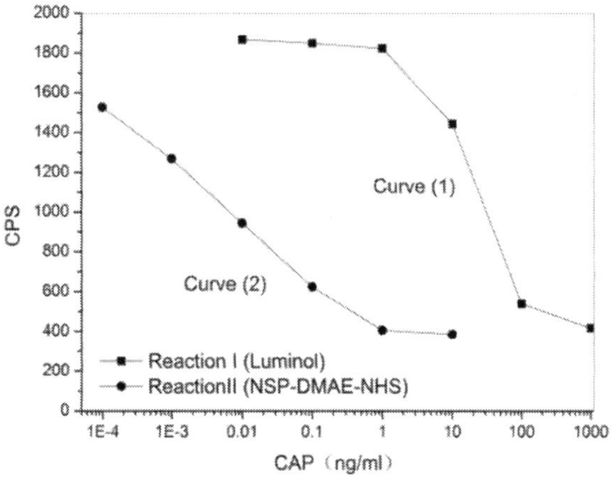

Figure.4. Calibration curves for detection of CAP in milk

3 Conclusions

In Fig.4, curve (1) showed the relationship of chemiluminescence intensity (CPS) versus the logarithm scale of concentrations of CAP succinate sodium in the range of 10^{-2}-10^{3} ng/ml. The curve (2) showed the CPS versus the logarithm scale of concentrations of CAP succinate sodium in the range of 10^{-4}-10 ng/ml. The IC_{50} values (the competitor concentration that causes 50% growth inhibition) of reaction I and reaction II were about 30 ng/ml and 0.03 ng/ml. The LOD value of reaction I and reaction II were approximately as 4 ng/ml and 0.008 ng/ml. In view of IC_{50}, the sensitivity of reaction II was improved 1000-folds compared with that of reaction I. This results indicates that applying NSP-DMAE-NHS as the chemiluminescent reagent is more sensitive and has a much wider measurement range than luminol in the CLIA. These results demonstrated that Reaction II was much more sensitive for the detection of CAP than some previously reported methods such as ELISA (IC_{50} of 10.5 ng/ml and LOD of 0.2 ng/ml, [20]), the pretreatment-free immunochromatographic assay (LOD of 10 ng/ml, [21]) and the time-resolved fluoroimmunoassay (TRFIA) (LOD of 0.05 ng/g, [22]).

In conclusion, Reaction II which used NSP-DMAE-NHS-labeled CAP-rabbit IgG as a tracer performed higher sensitivity and wider linear measurement range than reaction I which employed luminol with a tracer of CAP-HRP. Based on this, We will carry on to apply our system and method to detect other residues in food, and to make this biosensor commercialized in the future.

Acknowledgement

This work was supported by the National Natural Science Foundation of China [Grant No. 81371642]; the 111 Project (B14040).

References

1. M. J. Kasten, Mayo. Clin. Proc. **74**, 825 (1999)
2. X. Z. Shi, A. B. Wu, S. L. Zheng, R. X. Li, D. B. Zhang, J. Chromatogra. B. **850**, 24 (2007).
3. I. S. Park, N. Kim, Clin. Chim. Acta. ***578***, 19 (2006)
4. A.Y. Kolosova, J.V. Samsonova, A. M. Egorov, Food. Agr. Immunol. **12,** 115 (2000)
5. J. D. Louvois, J. Antimicro. Chemoth. **9**, 253 (1982)
6. V. Cerkvenik-Fajs,. Biomed. Chromatogr. **20**, 985 (2006)
7. H. Y. Shen, H. L. Jiang, Anal. Chim. Acta, **535**, 33 (2005)
8. M. X. Xie, Y.Liu, Y. M. Qiu, J. Han, Y. Z. Liu, Chinese. J. Anal. Chem. **33**, 1 (2005)
9. K. Fujita, H. Ito, M. Nakamura, M. Watai, M. Taniguchi, J-AOAC **91**, 1103 (2008)
10. S. B.Turnipseed, J. E. Roybal, A. P. Pfenning, P. J. Kijak, Anal. Chim. Acta. **483**, 373 (2003)
11. D. Arnold, A. Somogyi, J-AO*A*C **68**, 984 (1985)
12. V.Gaudin, N.Cadieu, P. Maris, Food. Agr. Immunol. **15**, 143 (2003)
13. N.Liu, S.Q. Song, L. Lu, D.X. Nie, Z. Han, X.L. Yang, Food. Agr. Immunol. **25**, 523 (2014)
14. S. Lin, S.Q. Han, Y.B. Liu, W.G. Xu, Anal. Bioanal. Chem. **382**, 1250 (2005)
15. X.Q. Tao, W. J. Wang, Z. H. Wang, X. Y. Cao, Luminescence **29**, 301 (2014)
16. T. Laube, S. V. Kergaravat, S. N. Fabiano, S. R. Hernandez, Biosens. Bioelectron. **27**, 46 (2011)
17. E. Zacco, M. I. Pividori, S. Alegret, Anal. Chem. **78**, 1780 (2006)
18. C. L. Xu, C. F. Peng, K. Hao, Z.Y. Jin, W. K. Wang, Luminescence **21**,126 (2006)
19. I. Weeks, I. Beheshti, F. McCapra, A. K. Campbell, Clin. Chem. **29**, 1474 (1983)
20. N. Sai, Y. P. Chen, N. Liu, G. G. Yu, P. Su, Y. Feng, Z. J. Zhou, *Talanta* **82**, 1113 (2010)
21. N. A. Byzova, E. A. Zvereva, A. V. Zherdev, S. A. Eremin, Talanta **81**, 843 (2010)
22. J. Z. Shen, Z. Zhang, Y. Yao, W. M. Shi, Y.B. Liu, Anal. Chim. Acta. **575**, 262 (2006)

Optimization of Composite Material Barrel Based on BP Neural Network Approximate Analysis

Xu Yadong [1,a], Li Yao [1] and Chen Yu [1]

[1] *Nanjing University of Science and Technology, Nanjing 210094, China*

Abstract. In order to improve the computational efficiency during the simulation and optimization process for the composite structure, the approximate analysis method was studied. The approximate analysis model was established for the composite material barrel by using BP neural network model. According to the network initialization and training, the model was used to replace the finite element analysis in the optimization process. The results showed that the BP neural network approximate model was effective and reliable, and the optimization efficiency was improved obviously.

1 Introduction

Composite materials can be designed. In order to design a composite structure with excellent performance in engineering, the structure performance analysis and optimization is needed. However, the actual engineering structure exists the following problems: Firstly, the model should be close to the physical model to obtain more accurate results, therefore, the complexity of the calculation model is high, the calculation has to occupy more computing resources and costs; Secondly, during the optimization design process, the optimization iteration needs to carry on a model analysis and to obtain the global optimal solution every time. That takes a long time to calculate. For example, In the process of adopting genetic algorithm combined with finite element analysis for structure optimization, in order to keep the diversity of population, the large population size is usually needed. The finite element program is needed to calculate each time, which makes the process is very slow.

In order to improve the computational efficiency and to speed up the optimization process, the structure approximate analysis technique was often used to replace the complex model so as to reduce the number of structure reanalysis. The artificial neural network method is a common method in structure analysis, which has a high degree of nonlinear mapping ability. Amounts of literature[1-5] showed that the method had been proved to be effective and reliable in the analysis of engineering structure.

In this paper, according to the optimization design of the composite material barrel of artillery weapon, the approximate model of the structure of the composite material barrel model was established by using BP neural network. And in the optimization process, the approximate analysis was used to replace the finite element analysis to improve the optimization efficiency.

[a] Corresponding author : dillxu@139.com

2 Composite structure and optimization model

During the firing process of artillery, the inner wall of the barrel is subjected to high temperature and high pressure gunpowder gas, and the common composite material is difficult to be subjected to this special environment. Therefore, the practical application of composite materials in the pipe is generally with a metal lining. At the same time, considering the actual situation of the processing technology and assembly of the composite barrel, the composite field has some limitations. Fig. 1 is the Schematic of composite material barrel composite position. The composite method is for the front part composite. Fig. 2 is the schematic of a composite part in the barrel. The metal lining is the thickness of the cylinder structure. Before and after the thickness difference of dh, the composite L is the length of the composite region.R0is the inner diameter of the barrel.h1, h2, h3 and h4are the thickness of the metal lining layer and the thickness of the three layer composite Respectively. The total thickness is the constant value.

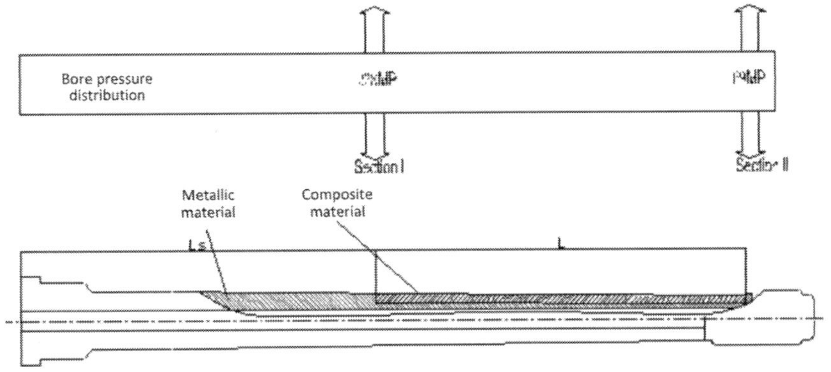

Figure 1. Schematic of composite material barrel composite position.

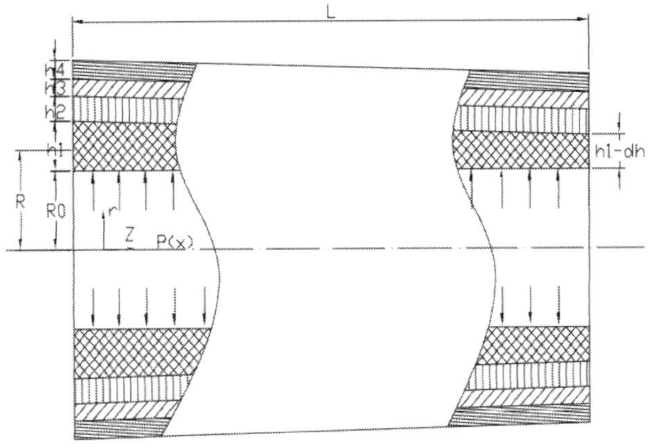

Figure 2. The schematic of the composite part of composite barrel.

According to the above composite structure model, four design variables h1、h2、h3and α for the composite barrel structure were selected as the optimal design parameters. α is the winding angle of the intermediate layer. The inner layer and outer layer of the composite material winding angle were 90° and 0° respectively. The natural frequency of f was as the target parameter for measuring the stiffness performance. The strength and weight of the barrel were as the constraint conditions. So the mathematical model of the optimization of the stiffness for the composite barrel was as follows.

$$
\begin{aligned}
& Maximize: f^*(X) \\
& \quad X=(x_1,x_2,x_3,x_4)^T=(h_1,h_2,h_3,\alpha)^T \\
& Subject \;\; to: \\
& \quad h_1+h_2+h_3<H \quad (h_1,\; h_2,\; h_3>0), \\
& \quad 0^{\circ}\le\alpha\le 90^{\circ}, C_1^*(X)\ge 1.224, w^*(X)\le\beta w_0
\end{aligned} \tag{1}
$$

In Eq.(1), f^* is the first order natural frequency of the composite barrel. C_1^* is the safety factor. w^* is the composite barrel quality.w0 is the quality of the whole metal barrel. β is the weight loss coefficient. H is the total wall thickness. f^*, C_1^* and w^* are the results respectively obtained by the calculation of the approximate model.

3 BP neural network approximation analysis model

3.1The network structure and parameters

BP neural network is the multilayer forward network which uses the error back propagation algorithm to train the weights of the network. The network structure and parameters include: the input / output node of the network, the number of layers, the activation function of each layer and the number of nodes in the hidden layer. In this paper, the BP neural network model was used to simulate the finite element calculation of the structure of the barrel. Its basic network structure was as shown in Fig. 3.

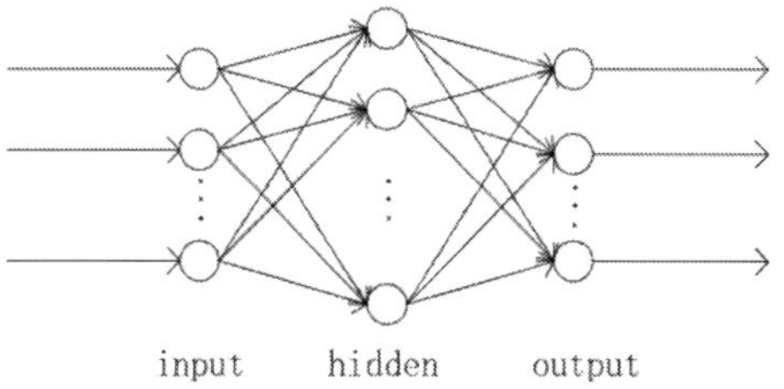

Figure 3. Structure of basic network.

The parameters were set as follows:

(1)Input nodes and output nodes

According to the above analysis, the total number of the composite barrel structure design parameters was four. The performance evaluation index number was three. Therefore, when designing a single multi-input and multi-output network, the input node number is 4, output node number 3. When designing a multi-input and single output network, namely the output parameters are designed to a single network individually. The number of output node is 4. The number of output node is 1.In order to compare the networks, two kinds of networks were designed, a single output network named BPN-II and a multi-output network named BPN-I. The BPN-I included the strength network, stiffness network and weight network.

(2)The number of plies

The kolmogorov multilayer neural network mapping existence theorem showed that the network concluding a hidden layer could approximate any nonlinear function. For the finite element calculation of the elastic structures, the continuity of the input and output mapping had been proved according to the elastic structure mechanics. Therefore, in order to make the network simple and easy to train, the network of only one hidden layer was selected while satisfying the requirements.

(3) The hidden layer nodes

The number of nodes in the hidden layer is the main factor that affects the performance of the network. The network approximation ability is weak when the number of nodes is too small. So the training process may not converge. There will appears over-fitting when the number of nodes is too more. The fault tolerance of the network will be lowered. The ability of "generalization" will be reduced. In this paper, by using the network reduction method, the hidden layer nodes of large numbers are set firstly, and then the number was reduced gradually. Ultimately, the 5 nodes were chosen according to the training of the network.

(4)The activation function

The nonlinear approximation capability of BP neural network is the activation function reflected by the tangent type S. The implicit layer selects the S type activation function generally, while the output layer uses the linear activation function generally.

3.2 The network initialization and network training

(1)The network initialization

Before the network training, the initial value of weight and threshold value are set firstly. In order to ensure that the network is not saturated with large weights, the weight of each layer w and threshold b are generally set to be the small random number.

(2) Collection of learning samples

The learning sample collection is essential after determining the structure and parameters of the network. The quantity and quality of the learning sample decide the network generalization ability in a large extent. Generally speaking, learning sample should satisfy compactness, ergodicity and fault tolerance. The learning samples collected by Latin square experimental design method can satisfy these conditions.

(3)The training process of BP network

In numerous training algorithm of BP network training, the Levenberg-Marquardt algorithm has a very fast convergence speed for the small and medium-sized network. The LM algorithm process is shown in Fig. 4. In the training parameters, the expected minimum error was taken as 0.0001, the maximum number of cycles was 5000, and the learning rate was 0.01 ~ 0.7.

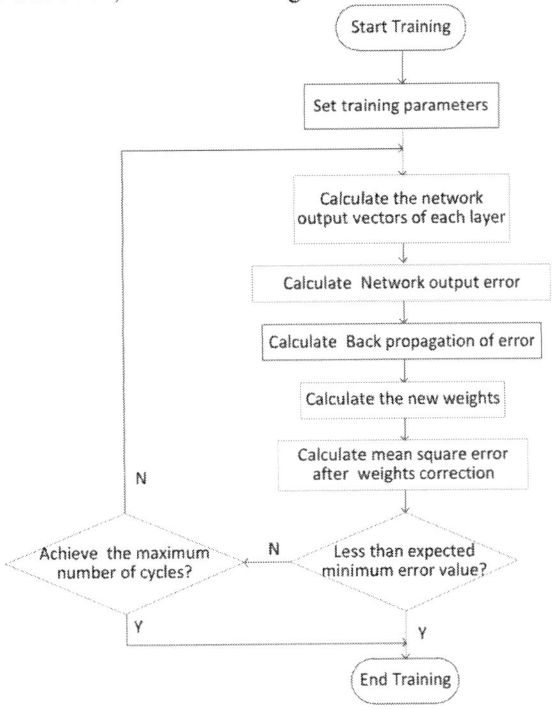

Figure 4. The training process of BP network .

The above parameters and the network training methods were used to train the 4 BP neural networks which have been established.

4 The optimal solution and results

By using BP neural network instead of finite element calculation, the computational efficiency is greatly improved, and the global optimization algorithm is adopted in the combination of approximate model for the optimization of composite structure. Therefore, in this paper, based on the BP neural network (BPN) model, the multi Island genetic algorithm (MIGA) is selected as the main optimization algorithm.

In order to further improve the optimization index after the end of the genetic evolution, combined with the local fast optimization algorithm - Sequential Quadratic Programming (SQP), the formation of the "MIGA - SQP - BPN" comprehensive optimization strategy, The algorithm flow chart is shown in Fig. 5.

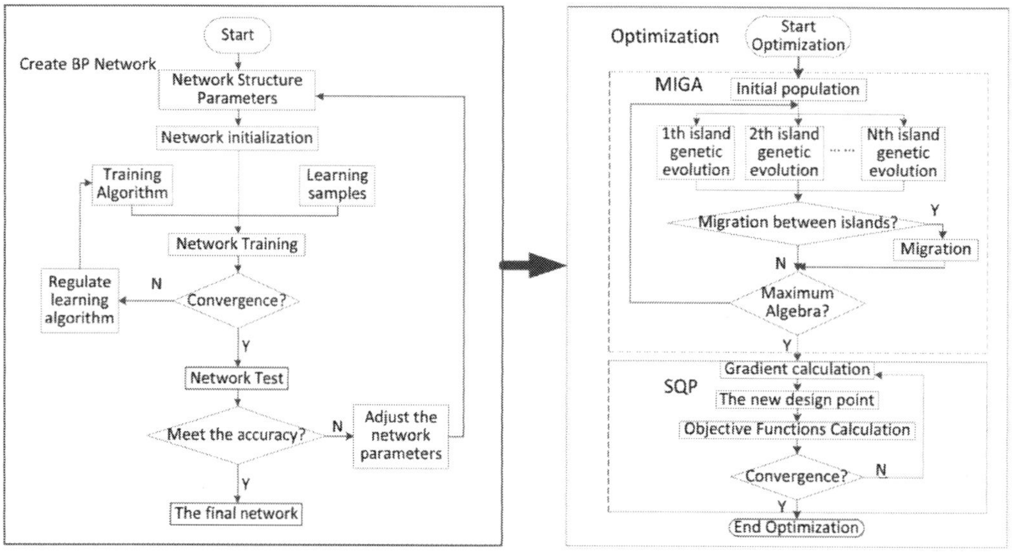

Figure 5. "MIGA - SQP - BPN" comprehensive optimization strategy flow chart.

Multi-island genetic algorithm is a pseudo-parallel genetic algorithm, comparing with traditional genetic algorithm, multi-Island genetic algorithm divides the entire evolutionary population into several subgroups, known as the "island", in each island subpopulations are independent of the selection, crossover and mutation operation regularly, in various islands some randomly selected individuals "migration" operation, will be transferred to the other islands, this way can maintain the diversity of the population, and inhibition of prematurity. Sequential quadratic programming is a mathematical programming method. The basic idea is using Lagrange function for quadratic approximation, generating a sequential quadratic programming sub-problems, sequential quadratic programming by forward difference and the center difference method to calculate the gradient of the objective function. Its main feature is the ability to quickly reach local minima in the local space for given initial solution close. The multi-island genetic algorithm and sequential quadratic programming combining comprehensive optimization, make full use of the complementarity of the two algorithms, multi-Island genetic algorithm to search in the global scope, when close to or reach optimal region, use the sequence quadratic programming to accelerate the local searching speed, making optimization process can quickly reach the optimum design point.

Using "BPN-MIGA-SQP" composite material barrel optimization involves two procedures. First step, creating composite material barrel BP neural network, according to the study sample and the training algorithm to train the network, through the network test, forming network; the second step,

 DOI: 10.1051/matecconf/20178802008

select genetic parameters, a global genetic evolution by MIGA, SQP local search after the end of the evolution. Finally, the optimal solution is obtained.

According to the optimization model and optimization strategy, the framework of composite material barrel optimization is established, as shown in Fig. 6.The preparation of optimization procedures, integrated BP neural network computing program. To optimize the solution, the MIGA algorithm and the SQP algorithm basic parameters are shown in Table 1.

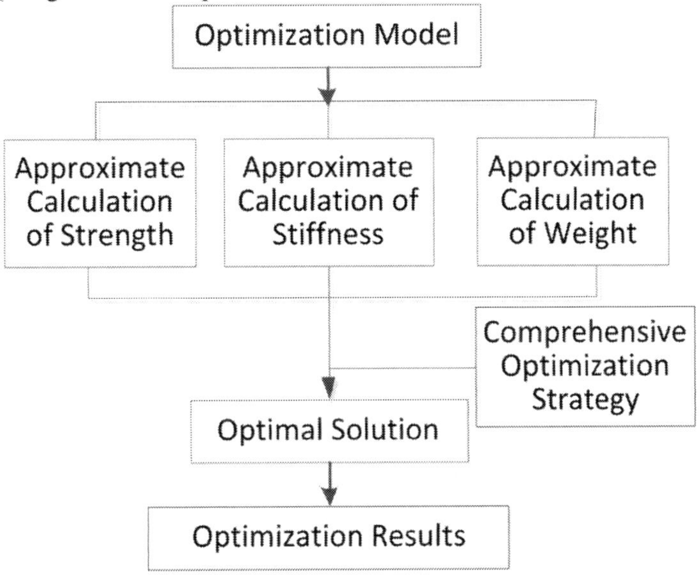

Figure 6. Optimization solving framework.

Table 1. Table of basic parameters of MIGA algorithm and SQP algorithm.

Max algebra	Number of subgroups	Number of individuals for subgroups	Crossover probability	Mutation probability	Migration probability
200	4	10	0.7	0.05	0.5
SQP Maximum number of iterations		SQP Iteration step		SQP Iterative termination	
100		0.001		0.00001	

Through optimization solution obtained composite material barrel optimal solution. Table 2 gives the comparison of the optimal solution obtained by the approximate model and the optimal solution obtained by the traditional mathematical programming method. In order to show the reliability of the approximate model, the finite element calculation results of the optimal scheme are given in this paper.

Table 2. The design schemes of optimization.

	h_1 (mm)	h_2 (mm)	h_3 (mm)	α (degree)	f_{max} (Hz)
Original optimal scheme	18.33	2.02	3.28	4.07	67.80
Latest program	17.39	1.09	4.99	3.76	69.56 (69.83)

By comparison, the optimal solution obtained by "MIGA - SQP - BPN" comprehensive optimization strategy was better than that obtained by the traditional mathematical programming method. The first-order natural frequency than the original optimization scheme improved the 3.36% and stiffness properties got better; through the comparison of the approximate calculation and finite element calculation of new scheme, the relative error between them was very small, which indicated that the validity and reliability of the BP neural network approximation model of this composite barrel.

5 conclusions

In this paper, the BP neural network is used to establish the approximate model of the composite barrel structure analysis, and it is applied to the structure optimization of composite material barrel. The results showed that the approximate analysis model based on BP neural network has high approximation accuracy. Due to the adoption of the structure approximation analysis and global optimization algorithm, the optimization efficiency is greatly improved, and the result is more ideal.

References

1. Cheng Yanxue, Pang Yongjie, Yang Zhuoyi, Research on composite material pressure hulls based on approximation, Ship Engineering, 37(2015)74-78
2. Wang Lei, Lu Jingui, Zhang Jiande, Hua Qi, Numerical empirical study on structural approximation analsysis of neural network for blade of wind turbine, Journal of Vibration, Measurement & Diagnosis,35(2015)116-119
3. Sei-Ichiro Sakata, Fumihiro Ashida, Masaru Zako, Approximates structural optimization using kriging method and digital modeling technique considering noise in sampling data, Computers & Structures, 86(2008)1477-1485
4. Olaf Amdt, Thomas Barth, Bemd Freisleben, Manfred Grauer, Approximating a finite element model by neural network prediction for facility optimization in groundwater engineering, European Journal of Operational Research, 166(2005)769-781
5. Olaf Amdt, Thomas Barth, Bemd Freisleben, Manfred Grauer, Approximating a finite element model by neural network prediction for facility optimization in groundwater engineering, European Journal of Operational Research, 166(2005)769-781

Effect of PBAT on Property of PLA/PHB Film Used for Fruits and Vegetables

Ma Xiuyu[1], Wang Yufeng[1,a], Wang Jianqing[1] and Xu Yaning[1]

[1]*College of Packaging and Printing Engineering, Tianjin University of Science & Technology, Tianjin, China*

Abstract: A blend of poly(lactic acid) (PLA) (70% by weight) and poly(3-hydroxybutyrate) (PHB) (30% by weight) with a solubilizing agent (PBAT) at four different concentrations (5%, 10%, 15% and 20% by weight per 100 parts of the blends) were investigated by FTIR, SEM, mechanical testing, water vapor permeability test and biodegradation studies. SEM showed that PBAT can improve the compatibility between PLA and PHB. The film show a great Oxygen transmission rate and Water vapor permeability property when the content of PBAT is 10%.

1 Introduction

In recent years, sustainability, environmental concerns and green chemistry have played a big role in guiding the development of the next generation of materials, products and processes. The persistence of plastics in the environment, dwindling petroleum resources, shortage of landfill space and the concerns over emissions of toxic gases during incineration have fuelled efforts to develop biodegradable polymers from renewable resources[1]. The high social demand for sustainable biomaterials in the production of consumer goods has lead to increased attention in "green" technologies, in particular by developing environmentally friendly packaging systems [2, 3]. The proposal of new structures coupling adequate functionalities and sustainability is one of the current challenges in the research for new packaging materials. Important efforts have been recently devoted to obtain high performance bio-based materials, including polymer matrices, nanostructures and additives, fully derived from renewable resources [4, 5].

Poly(lactic acid) (PLA) is a linear thermoplastic polyester derived from the fermentation of renewable agricultural crops, such as starch, cellulose and other polysaccharides [6]. Moreover, it is compostable disappearing completely in less than one month [7, 8] and it is currently the most used biobased and biodegradable polymer for short-term applications. However, the main drawbacks of PLA are that it shows lower thermal and mechanical performance with respect of their petrochemical counterparts. Poly (3-hydroxybutyrate) (PHB), an aliphatic polyesters, synthesized by microorganisms, with high crystallinity and high melting point, is often used as components for blending with PLA, and leads to materials with interesting physical, thermal, and mechanical properties compared to neat PLA. This paper focuses on the effect of on oxygen and water vapor permeability of PLA/PHB film through using PBAT as a solubilizing agent, in order to find the best amount of additives, and explore

[a] Corresponding author : ppcwyf@tust.edu.cn

the controlled conditions of fruits and vegetables film properties.

2 Materials and methods

2.1 Materials

The PLA (4032D, NatureWorks Co. LLC, USA) had an average molecular weight of 52,000 g·mol^{-1}, a polydispersity (M_w/M_n) of 1.9 and a high optical purity with 98% L-lactide content. The poly(3-hydroxybutyrate) (PHB under the trade name EM10080) was purchased from Shenzhen Ecomann Biotechnology Co., Ltd. The poly(butylene adipate-co-terephthalate) (PBAT ecoflex® F Blend C1200) was obtained from BASF.

2.2 Preparation of the blends

The materials were pre-treated by drying at 60℃ for 24h to eliminate possible absorbed water on the surface of the particles. Compositions containing 70% PLA and 30% PHB in weight were mixed, respectively designated PBAT at four different concentrations (5%, 10%, 15% and 20% by weight per 100 parts of the blends). All the formulations were melt-compounded in a single screw extruder. The mixture was granulated after melt blending extrusion within the torque rheometer (RM-200A, Harbin HARP electrical technology Co., Ltd). The melt compounding temperature at different zones was independently controlled in order to achieve a temperature profile in the range of 170~190℃. The screw speed was set at 25RPM.

2.3 Morphological characterization by scanning electron microscopy (SEM)

Blend materials were fractured in liquid nitrogen to observe the interior of the unstressed composite. A sample was dropped directly into liquid nitrogen and fractured with a pre-chilled razor blade held in a vice-grip. The fractured pieces were picked out of the liquid nitrogen using pre-chilled forceps and placed in a desiccator to thaw and reduce the condensation of water on the surface of the material. All specimens were coated with gold-palladium for 45s. Specimens were viewed in a Hitachi S4700 field emission scanning electron microscope (Hitachi HTA, Japan) at 2 kV.

2.4 Mechanical property

An electronic universal material testing machine was used to measure mechanical property of films. And the referring standard of measurement was GB/T13022-91—Plastics-Determination of tensile properties of films. The reported values were the average of at least 5 measurements.

2.5 Fourier transformed infrared spectroscopy (FTIR)

Infrared spectroscopy analysis (FT-IR) was performed at room temperature in transmission and reflection modes by using a VECTOR 22 spectrometer at a wavenumber range 4000-400 cm^{-1}.

2.6 Oxygen transmission rate (OTR) and Water vapor permeability

The OTR was measured by using a GDP-C oxygen permeation analyzer (Brugger, Germany) at room temperature. 12cm×12cm films were prepared referring to GB/T 1038-2000'Plastics-Film and sheeting-Determination of gas transmission Differential-pressure method'. WVTR of film sample was determined gravimetrically at 38℃ and 90% RH conditions using water vapor transmission measuring cups in accordance with the GB/T 1037-88 'Test method for water vapor transmission of plastic film and sheet-Cup method' standard method with a modification. Circle sample was sheared about 65 mm in diameter, and then put above the desiccant in moisture vapor transmission cup, fixed by sealing wax.

2.7 Soil burial test

Biodegradation studies under real soil burial conditions were carried out within the campus of Tianjin University of Science & Technology, Tianjin, China. The soil biodegradation test lasted for two months. The film samples were cut into 3cm×4cm pieces and all the samples were buried 10-20cm under the soil. Every ten day a sample was retrieved, and the corresponding film samples were cleaned and dried at room temperature. The film weight before and after soil biodegradation test was recorded accordingly. Then calculate the weight loss rate.

3 Results and discussion

3.1 Mechanical properties

The films intended for food packaging applications are subjected to elongation during the packaging procedure, being important their elastic modulus and elongation characteristics. Tensile strength and elongation at break properties of different films are represented in Table 1. The film of PLA/PHB without PBAT has a very high tensile strength and a low elongation at break. Because the composite film contain 70% of PLA which is a glossy polymer. The blends of PLA/PHB with PBAT showed different behavior. The tensile strength was significantly reduced by addition of PBAT, but the elongation at break increased in value. PLA/PHB without PBAT had a higher Young's modulus (YM) than that for other material blend with PBAT. The introduction of PBAT into the system provoked a significant reduction in Young modulus and tensile strength.

Table 1. Effect of PBAT content on mechanical properties of film

Code	Young's modulus (MPa)	Tensile Strength (MPa)	Elongation at break (%)
0	2374.94	36.36	318.42
5%	2235.05	32.40	339.04
10%	2126.48	31.68	369.51
15%	2155.19	28.98	372.03
20%	2071.43	25.61	382.49

3.2 Water vapor permeability (WVP) and Oxygen transmission rate (OTR)

The oxygen transmission permeability and water vapor permeability were studied and the obtained results are shown in Figure 1. The WVP of PLA/PHB without PBAT was 9.52×10^{-15}g·cm/cm^2·s·Pa. The WVP of PLA/PHB film increased by blending with PBAT. The value of WVP is increased quickly with the addition of PBAT first and then become steady. It is well known that plasticizers addition increases the free volume in PLA matrix and as a result a reduction in the oxygen barrier properties is produced. The oxygen transmission permeability of films increased with increase in PBAT content.

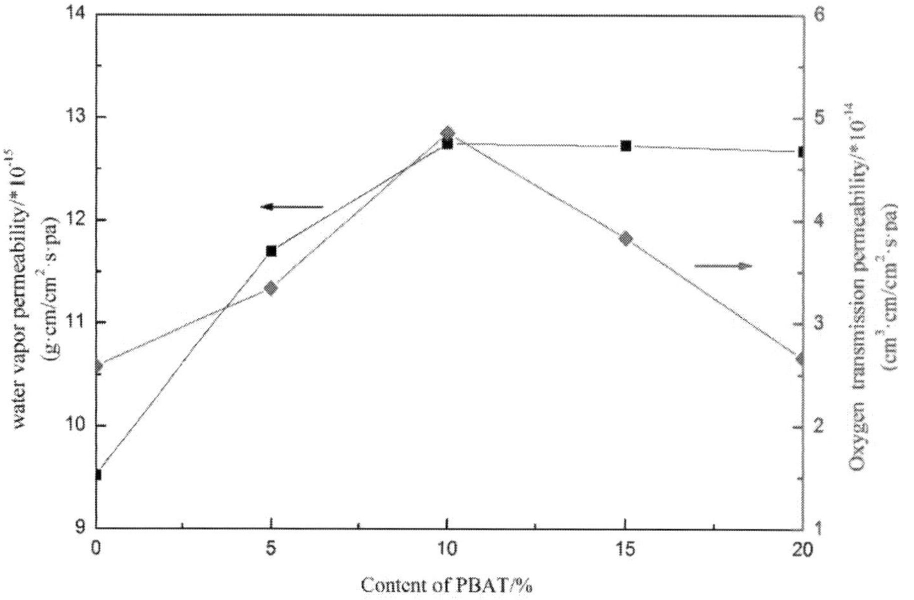

Figure 1. Effect of PBAT on oxygen and water vapor permeability of film

3.3 FT-IR analysis

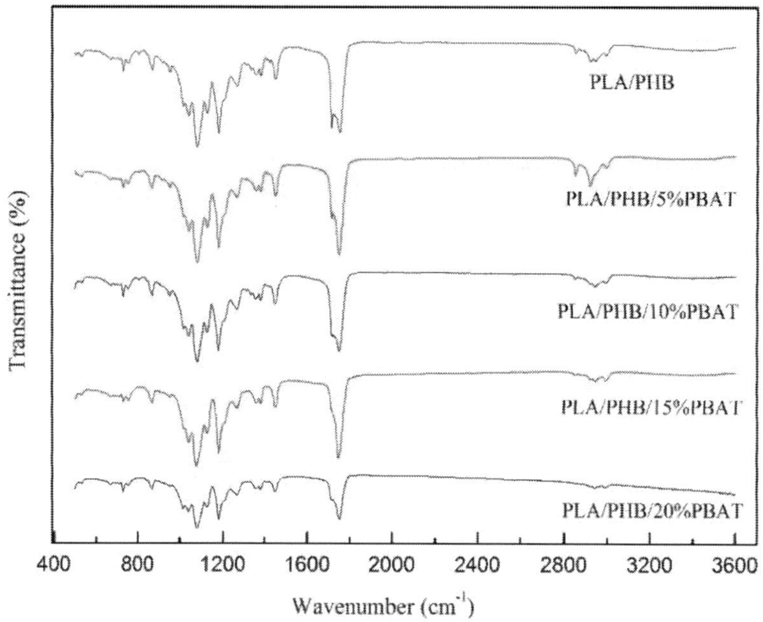

Figure 2. FT-IR spectra of PLA/PHB and PLA/PHB/PBAT samples in 400cm^{-1}~3600cm^{-1}.

FTIR spectra of PLA/PHB and PLA/PHB/PBAT films are shown in Figure 2 with an offset comparison. In the PLA film, the peak at the wavenumber of 3502 cm^{-1} represents CH_3 stretching vibration. The peaks at 2995 and 2945 cm^{-1} clearly are associated with the antisymmetric and symmetric stretching vibrations of CH_3 of saturated hydrocarbons, respectively [9]. At 1085 cm^{-1} a band associated with the –O–C– asymmetric mode of PLA appears [10]. Meanwhile, at 1180 cm^{-1} a band associated with the –C–O– bond stretching in the –CH–O– group of PLA appears [11]. There is a band at 1278 cm^{-1} assigned to the crystalline –C–O–C stretching of PHB and At 1360 cm^{-1} the CH deformation and asymmetric bands appear. The typical asymmetric stretching of the carbonyl group (–C=O) in neat PLA mat is centered at 1752 cm^{-1} and it is attributed to the amorphous carbonyl vibration [10]. It is possible to observe a small shoulder near 1740 cm^{-1} related with the amorphous state of PHB, showing that the PHB presents mainly crystalline structure with minor amorphous component [12]. In the PLA/PHB/PBAT blend films, peaks with low intensity which revealed the stretching vibration of C–O at 1718 cm^{-1}, the symmetric stretching vibration of C–O at 1271 cm^{-1}, and the bending vibration absorption of CH- plane of the benzene ring at 729 cm^{-1} of PBAT were found in addition to the peaks of PLA. The peaks positions of PBAT and PLA did not change after blending, which indicates that there was no chemical interaction between these two polymers [13].

3.4 Morphological properties

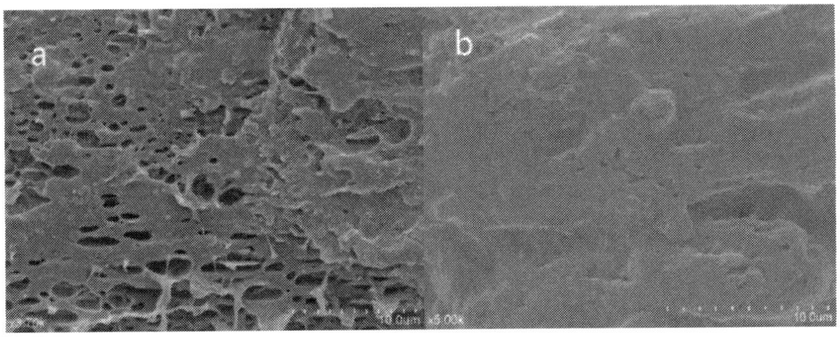

Figure 3. SEM images of PLA/PHB (a) and PLA/PHB/PBAT (b)

The initial particle size, the polymer elasticity, the dispersed phase percentage and the draw ratio are the main factors affecting the morphology formed during drawing. The morphologies of the fracture surfaces were investigated by SEM, as shown in Fig.3. The PLA/PHB film without PBAT presents an irregular fracture surface due to the crystalline structure of PHB in blends. There is almost no hole at the fracture surface in Fig. b, in which the PBAT fill in it.

3.5 Biodegradation testing

From the weight loss data in Table 2, all of the samples have lost certain amount of weight after the soil biodegradation test. For instance, after the first ten days, the PLA/PHB film lose the weight of 0.8%, while the weight of others were 1.08%, 1.14%, 1.06% and 2.14%, respectively. The weight loss of PLA/PHB blends with PBAT are all higher than that of PLA/PHB, which indicate that PBAT can increase the degradation rate. Also, the degradation rate is relate to the content of PBAT in blends.

Table 2. Film weight loss after the soil burial test

Entries	Weight loss rate (wt%)					
	10d	20d	30d	40d	50d	60d
0	0.8	1.24	2.86	4.15	5.96	7.35
5	1.08	2.47	4.01	6.58	7.35	9.66
10	1.14	2.45	4.46	6.59	7.86	9.97
15	1.06	2.68	4.85	6.89	7.83	10.56
20	2.14	3.26	5.65	7.74	8.96	11.53

DOI: 10.1051/matecconf/20178802009

4 Conclusions

The current study investigated the PBAT effect on the morphology, mechanical properties, as well as on the WVP and OTR properties, of PLA//PHB/PBAT blends in the form of blown films. Mechanical properties test showed that PLA/PHB blends with PBAT could increase elongation at break and introduction of PBAT into the system provoked a significant reduction in Young modulus and tensile strength. SEM results showed that PBAT can improve the compatibility between PLA and PHB. The soil biodegradation test proved that the weight loss increased with the content of PBAT increasing.

References

1. Mohamed A., Abdelwahab, Allison Flynn, et al. Thermal, mechanical and morphological characterization of plasticized PLA/PHB blends [J]. Polymer Degradation and Stability, 2012, 97: 1822-1828.
2. Murali M. Reddy, Singaravelu Vivekanandhan, Manjusri Misra, et al. Biobased plastics and bionanocomposites: current status and future opportunities [J]. Progress in Polymer Science, 2013, 38 (10): 1653-1689.
3. Perlette N. Takala, Stephane Salmieri, Afia Boumail, et al. Antimicrobial effect and physicochemical properties of bioactive trilayer polycaprolactone/methylcellulose-based films on the growth of foodborne pathogens and total microbiota in fresh broccoli [J]. Journal of Food Engineering, 2013, 116 (3): 648-655.
4. Arrieta, M. P., Peltzer, M. A., López, J., et al. Functional properties of sodium and calcium caseinate antimicrobial active films containing carvacrol [J]. Journal of Food Engineering, 2014, 121: 94-101.
5. Arrieta, M. P., Samper, M. D., López, J., et al. Combined effect of poly (hydroxybutyrate) and plasticizers on polylactic acid properties for film intended for food packaging [J]. Journal of Polymers and the Environment, 2014, 22 (4): 460-470.
6. M.P. Arrieta, J. López, D. López, et al. Biodegradable electrospun bionanocomposite fibers based on plasticized PLA–PHB blends reinforced with cellulose nanocrystals [J]. Industrial Crops and Products, 2016, 93: 290-301.
7. Arrieta, M.P., López, J., Rayón, E., et al. Disintegrability under composting conditions of plasticized PLA-PHB blends [J]. polymer degradation and stability, 2014, 108: 307-318.
8. Dolores, S.M., Patricia, A.M., Santiago, F., et al. Influence of biodegradable materials in the recycled polystyrene [J]. Journal of Applied Polymer Science, 2014, 131.
9. Al-Itry, Racha, Lamnawar, Khalid, Maazouz, Abderrahim. Rheological, morphological, and interfacial properties of compatibilized PLA/PBAT blends [J]. Rheologica Acta, 2014, 53 (7): 501-517.
10. R. Auras, B. Harte, S. Selke. An overview of polylactides as packaging materials [J]. Macromolecular Bioscience, 2004, 4 (9): 835-864.
11. E. Fortunati, D. Puglia, C. Santulli, et al. Biodegradation of Phormium tenax/poly(lactic acid) composites [J]. Journal of Applied Polymer Science, 2012, 125 (2): E562-E572.
12. T. Furukawa, H. Sato, R. Murakami, et al. Structure, dispersibility, and crystallinity of poly(hydroxybutyrate)/poly(L-lacticacid) blends studied by FT-IR microspectroscopy and differential scanning calorimetry[J]. Macromolecules, 2005, 38 (15): 6445-6454.
13. Long-Feng Wang, Jong-Whan Rhim, Seok-In Hong. Preparation of poly(lactide)/poly(butylene adipate-co-terephthalate) blend films using a solvent casting method and their food packaging application [J]. LWT-Food Science and Technology2016, 68: 454-461.

DOI: 10.1051/matecconf/20178803001

Experimental Studies on Improved Plastic Bonded Explosives Materials (PBXs) for Controlled Fragmentation Warheads

Karim Elsharkawy[1,a], Lin Guo[1] and Hany Fouda[1]

[1]*School of Chemistry and Environment, Beijing University of Aeronautics and Astronautics, Beijing 100191, China*

Abstract: This paper describes formulations of plastic bonded explosives (PBXs) based on three highly brisant explosives, namely 1,3,5-trinitro-1,3,5-triazinane (RDX), 1,3,5,7-tetranitro-1,3,5,7-tetrazocane (HMX) and pentaerythritoltetranitrate (PETN) with polyurethane (PU) based on Glycidyl azide polymer (GAP) as an energetic binder in comparison with composition-B, which used in the fragmentation warheads. The sensitivity and performance properties of different types of PBXs were evaluated by experimental results of prepared selected PBX compositions. Casting technique was used to prepare the selected compositions containing 14% PU based on GAP. It has been observed that the brisance of the PBX based on HMX was higher than that of comp-B by 21.3 %, the detonation velocity showed a remarkable increase of the order of 8480 (m/s) while that of comp-B was 7638 (m/s). A controlled fragmentation warhead with an outer grooving warhead case of dimensions 100x35x4 mm was used and arena test was carried out to determine the lethal zone of the fragmentation warhead. The lethal zone obtained from arena test for PBX composition based on HMX named PBXHG4 was higher than that based on RDX or PETN, and than that of comp-B by 40%.

1 Introduction

The new development in the field of explosives is to develop a new class of explosives known as plastic bonded explosives (PBXs) which have significantly lower vulnerability to various stimuli than conventional high explosives as comp-B that suffers from various disadvantages such as relatively poor impact sensitivity, cure shrinkage which results in voids and cracks, and somewhat violent reaction during cook-off [1-3]. By the end of the First World War, the explosives development takes the direction to improve the explosive performance, and as a result, the sensitivity of explosives has been increased. Over the years, many accidents were happened during handling, transporting, testing and manufacturing of explosives, where many researchers have always paid attention to increase performance without considering safety aspect. The demand for increased safety in explosives transportation, handling, and storage has led to the improvement of Insensitive Munitions (IM) [4]. The design of these weapons and explosives decreases the probability of unwanted and unexpected detonation from external stimuli such as weapon fragments, heat, and shock. This can be achieved by the modification to the explosive formulation.

Plastic bonded explosive (PBXs) are various explosive mixtures where the crystals of high explosives are embedded into a thin layer of polymeric material. Recently, many researches focused on the preparation, characterization and applications of modern of PBX formulations based on high

[a] Corresponding author: karimelsharkawy@buaa.edu.cn

brisance explosives and energetic polymeric binders to achieve better explosive properties and low vulnerability [5-8]. Controlled fragmentation warheads are used to defeat almost all types of targets (underwater, aerial, human and underground) to set them out of action by the action of fragments created from the warhead body at the time of explosion. The velocity of fragments depends mainly on the detonation velocity or brisance of the high explosive. Higher fragment velocity and large lethal zone could be obtained by using high performance explosive material [9,10].

2 Experimental

2.1 Preparation of Plastic bonded explosives

All the chemicals used in this work; RDX, HMX, PETN, GAP, and Isophoron diisocyanate (IPDI - crosslinking agent) were of high purity. These PBXs formulations based on polyurethane were prepared by using the casting technique under vacuum [11]. This process was conducted in a stainless steel bowl of 5 kg capacity provided with double wall jacket, where a circulating liquid could be either heated or cooled. Stainless steel bowl is equipped with a vertical mixer with three blades rotating in an orbital motion. The prepolymer used in this work was GAP (1.43 measured mg equivalent OH/g GAP). The prepolymer was precisely dropped into the reactor and then preheated until the temperature reaches to 60 °C then the dried pure explosive was divided into four equal portions and added during stirring without vacuum for 8 minutes for every portion. After adding the fourth portion of the explosive, the mixer was kept under vacuum for18 minutes to avoid the presence of air bubbles in the mixing samples. Finally, the curing agent (IPDI) was added and mixing was carried out for relatively short time. The explosive mixture was then poured in polyvinyl chloride (PVC) mould tubes and experimental controlled fragmentation warhead to measure detonation velocity and lethal zone respectively after curing which was carried out at 55-60 °C for 12 days. Three PBX formulations (86% explosive + 14% binder) were prepared at the same binder PU based on GAP, but at different explosive filler HMX, RDX, and PETN, which designated as PBXHG4, PBXRG4, and PBXPG4 respectively.

2.1 Evaluation of PBXs formulations

2.2.1 Sensitivity tests

Sensitivity to friction was determined using BAM friction test apparatus. The frictions test was determined by the percentage of initiation by changing the loading of the pistil [12]. Sensitivity to impact was carried out using IKa Maschinenbau apparatus, using 5 Kg falling weight [13]. Upper sensitivity limit was used to identify the minimal height at which 100% initiation was achieved. The sensitivity to heat was obtained by measuring the ignition temperature for the prepared PBX samples using Chilworth deflagration test apparatus [12]. To determine the ignition temperature, the temperature was uniformly increased [5 °C/min] until the explosion conversion occurred.

2.2.2 Detonation velocity

The prepared PBX samples were pressed to a density 1.6 (g/cm^3) in PVC tubes of 2 mm inside diameter and 190 mm length. The Exploment -Fo-2000- Multi Channel (Swiss made) was used to measure the detonation velocity of these formulations. The time interval for a detonation wave to travel a known distance between the two fiber optic probes in microseconds was displayed with the calculated detonation velocity in (m/s) [14].

 DOI: 10.1051/matecconf/20178803001

2.2.3 Brisance

This test was carried out using the brisance testing unit according to Kast technique [12], where 2.5 g of the explosive charge was pressed to a density of 1.6 (g/cm^3) into an aluminum tube of 30 mm height, 12 mm inside diameter and 4 mm wall thickness. Special test detonator with 0.2 g of lead azide and 0.4 g of RDX was prepared on site to initiate the explosive charge. The copper crushers used are of 9.8 mm height and 6 mm diameter. The deformation (final lengths) of copper crushers was determined after the explosion of the charges and converted into force units according to the calibration table of the copper crusher static compression force test [15].

2.2.4 Arena test

The arena test is used to measure the effects of high-explosive warhead detonations to determine the lethal zone. Lethal zone is one of the most important parameter of HE warhead efficiency [16]. The used controlled fragmentation warhead of dimensions (100 mm length, 35 mm diameter and 4 mm thickness) made from steel 35 as shown in Fig. 1, was selected to be filled with 100 g of the investigated PBX formulations. In this test, we compare lethal efficiencies for similar warheads; the warhead is put on the ground at which the warhead nose is in the down position in the center of the test arena. The target is made of metallic witness panels of dimensions (2 m height, 1 m width and 2 mm thickness) placed at specified distances (10, 15, 20, 25, 30 and 35 m) from tested warhead. Electric detonator was used to initiate the high explosive charge, after the detonation of the warhead, number of perforations at metallic witness sectors are counted per square meter for each sector. A characteristic distance is the fragments concentration density of one perforation per square meter. This distance is named warhead efficiency radius, which means that warhead with greater lethal zone has corresponding greater efficiency radius [17,18].

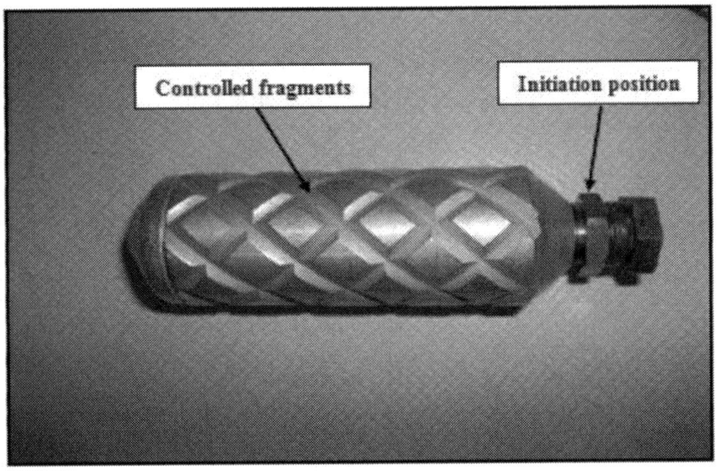

Figure 1. Photograph of the used controlled fragmentation testing cylinder casing.

3 Results and Discussion

3.1 Sensitivity tests

For comp-B (60% RDX + 40% TNT), and the prepared PBX formulations, the sensitivity results are listed in Table 1. It is clear that for PBXHG4, the decrease of sensitivity to impact was 42.5%

DOI: 10.1051/matecconf/20178803001

compared with comp-B. PBXRG4 and PBXHG4 showed no initiation even when applying the maximum friction force (360 N) of the test apparatus but PBXPG4 showed initiation because of PETN is high sensitive to friction. PBXs showed ignition temperature slightly lower than comp-B. This can be attributed to the fact that the coating of polyurethane based on GAP acts as heat sensitizing medium because its softening temperature is less than 152 °C. PBXHG4 has sensitivity to heat close to that of comp-B. The values of ignition temperature for the prepared PBX formulations and comp-B are listed in Table 1.

3.2 Detonation velocity

The detonation velocity of explosives is one of the important terms which represent the rate of energy delivery through explosive conversion. From the obtained results we can find that the type and content of binder have a significant effect on detonation velocity of the prepared PBX formulations. The values of detonation velocity (D) were determined for comp-B and the prepared PBX formulations are listed in Table 1. For PBXRG4, detonation velocity increased by 6.9% but increased by only 3% for PBXPG4. For PBXHG4, the increase of detonation velocity was 11%, when compared with that of comp-B.

3.3 Brisance

The values of brisance (B) were determined for comp-B, and the prepared PBX formulations are listed in Table 1. Brisance were slightly increased for PBXPG4 by 6.9 % and for PBXRG4 increased by 9.2%, for PBXHG4 the increase of brisance was 21.3%, when compared with that of comp-B, which mean a great improvement in shuttering effect, improved velocity of the fragment and the lethal zone.

Table 1. Sensitivity and performance results of comp-B and PBX formulations based on PU (GAP).

Explosive	Impact (J)	Friction (N)	Ignition temp. (°C)	Detonation velocity (m/s)	Brisance (kp)
Comp-B	12.0	220	239	7638	1059
PBXRG4	19.8	>360	232	8165	1157
PBXHG4	17.1	>360	234	8480	1285
PBXPG4	15.3	238	228	7867	1132

3.4 Arena test

The lethal zone is one of the most important parameters of high explosive warhead efficiency. The warhead lethality zone for comp-B and the prepared PBX formulations are shown in Fig. 2. Only one fragment is sufficient to kill or affect on the target so, a horizontal line was sketched at fragment density = 1 and the radius of lethality zone for each examined explosive charge was obtained from the X axis. It is clear that warhead filled with PBXHG4 has a greater lethal efficiency (28 m) than PBXRG4 (25 m), PBXPG4 (23 m) and comp-B (20 m). For PBXHG4 the increase of efficiency radius was 40%, when compared with that of comp-B. From Fig. 3, it is clear that fragments sizes were controlled as regular and effective from warhead casing to avoid its random distribution which could be achieved by grooving the outer surfaces of the warhead casing. After the explosion of the warhead; the explosion force affect on the weak positions in the warhead structures which are located between the formed fragments and produce a controlled fragment size and shape.

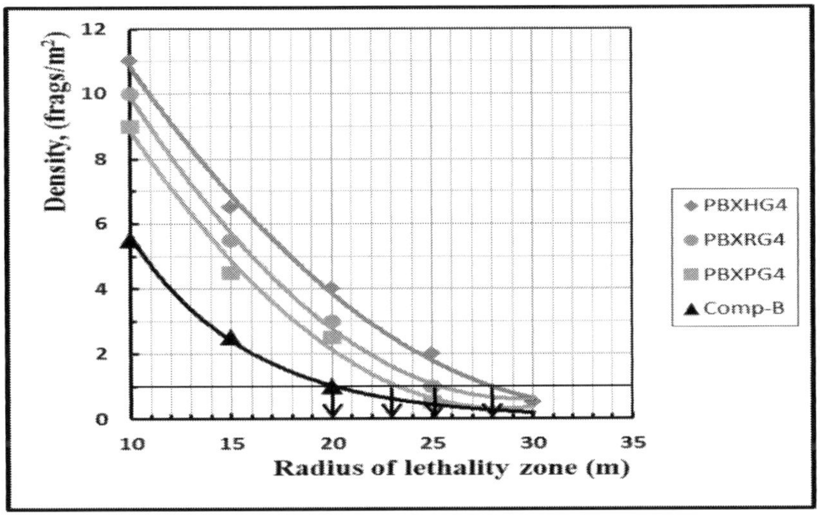

Figure 2. Warhead lethality zone of the warhead filled with comp-B and PBX formulations.

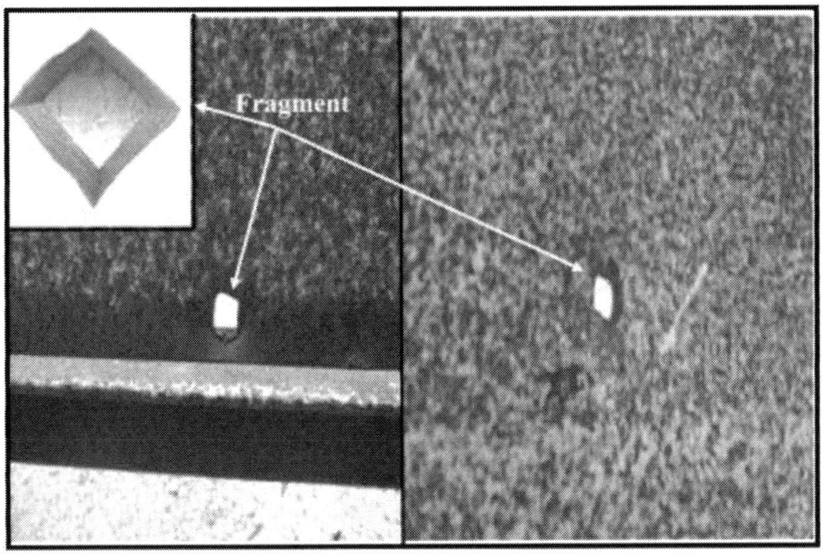

Figure 3. Photograph of the controlled fragment after the detonation of the warhead and its effect on the witness sectors.

4 Conclusions

In the present study, three PBX formulations based on 86% explosive material (RDX, HMX, and PETN) and 14% binder (PU based on GAP) have been formulated using casting technique and investigated for explosive and performance properties. The choice of this technique increases the safety during production. The measured sensitivity characteristics proved that PBXHG4 has lower sensitivity to the mechanical stimulus (impact, friction) than that of comp-B by 42.5% for impact sensitivity. Its sensitivity to heat is close to that of comp-B. PBXHG4 has higher performance parameters, in which detonation velocity increased by 11%, brisance increased by 21.3% and lethal zone increased by 40% when compared with that of comp-B. PBXs possess several advantages: ease and simplicity of production, lower sensitivity to impact and friction, suitable heat sensitivity, performance and increased safety. Thus, it can be said that these formulations especially PBXHG4 which have high values of brisance, are recommended to be used as a main explosive charge of low sensitive fragmentation warheads and demolition charges instead of composition-B.

References

1. T.Urbanski, *"Chemistry and Technology of explosives"* Pergamon press, New York, USA, **4** (1985)
2. Doll D. W., Hanks J. M., Niles J. H., "Insensitive Composition B Replacement" *Insensitive Munitions and Energetic Materials Technology Symposium*, Tampa, Florida, USA, **2**, 264 (1999)
3. M. M. Ahmed, M. H. Abuuznien, B. M. Sidig, " Replacement of Comp-B by Plastic Bonded Explosives In a Selected Warhead" *7th International Conference on Chemical & Environmental Engineering*, Cairo, Egypt, (2014)
4. A. Singh, M. Kumar, P. Soni, M. Singh, A. Srivastava, " Mechanical and Explosive Properties of Plastic Bonded Explosives Based on Mixture of HMX and TATB" Defence Science Journal, **6**, 622-629 (2013)
5. M. Radwan, "Explosive Characteristics of Aluminized PBXs Based on HMX and Polyurethane Binder", *32nd International Annual Conference of ICT*, Karlsruhe, Germany (2001)
6. I.Plaksin, J.Campos, R.Mendes "Shock sensitivity and detonation performance of PBX formulations based on the RS-RDX, RS-HMX and fine/ultra-fine HE particles bonded with the inert binders" *39th International Annual Conference of ICT*, Karlsruhe, Germany (2008)
7. M. H. Hamad, H. E. Mostafa, " Study of Advanced Plastic Bonded Explosives" *7th International Conference on Chemical & Environmental Engineering*, Cairo, Egypt, (2014)
8. Q. L. Yan, S. Zeman, A. Elbeih, A. Zbynek, " The Influence of the Semtex Matrix on the Thermal Behavior and Decomposition Kinetics of Cyclic Nitramines" Central European Journal of Energetic Materials, **10**, 509-528 (2013)
9. M. T. Shahraini, "Casing toughness effect on anti-air fragmentation warhead performance", *23rd International Symposium on Ballistics*, Tarragona, Spain (2007)
10. S.Waggener, "The evolution of air target warheads", *23rd International symposium on ballistics*, Tarragona, Spain (2007)
11. M. B. Talawar, S. K. Jangid, T. Nath, R. K. Sinha, S. N. Asthana" New directions in the science and technology of advanced sheet explosive formulations and the key energetic materials used in the processing of sheet explosives: Emerging trends" Central European Journal of Energetic Materials, **300**, 509-528 (2015)
12. M. Suceska, "*Test methods for explosives*", Springer, ISBN 0-387-94555-5 (1995)
13. "Recommendations on the transportation of dangerous goods, manual of tests and criteria", Third edition, United Nations, New York and Geneva (1999)
14. "Exploment – Fo – 2000 apparatus catalouge", Switzerland (2000)
15. "OZM Research brisance apparatus catalogue", Czech Republic (2010)
16. B.Zecevic, A.Catovic, J.Terzic,"Comparison of lethal zone characteristics of several natural fragmenting warheads", Central European Journal of Energetic Materials, **5**, 67-81(2008)
17. B.Zecevic, J.Terzic, A.Catovic, S.S.Kadic, "Influencing parameters on HE projectiles with natural fragmentation", *9th New Trends in Research of Energetic Materials Seminar* , Czech Republic (2006)
18. B.Zecevic, J.Terzic, F. Razic, S.S.Kadic, "*Lethal Influence Factors of Natural and Performed Fragmentation Projectiles*", DAAAM International Scientific Book, ISBN 978-3-902734-05-1, 219-234 (2015)

The Adsorption Geometry and Electronic Structure of Organic Dye Molecule on TiO_2(101) Surface from First Principles Calculations

Mang Niu[1,a]

[1]*College of Science, China University of Petroleum (East China), Qingdao 266580, Shandong Province, China*

Abstract. Using density functional theory (DFT), we have investigated the structural and electronic properties of dye-sensitized solar cells (DSSCs) comprised of I-doped anatase TiO_2(101) surface sensitized with NKX-2554 dye. The calculation results indicate that the cyanoacrylic acid anchoring group in NKX-2554 has a strong binding to the TiO_2(101) surface. The dissociative and bidentate bridging type was found to be the most favorable adsorption configuration. On the other hand, the incorporations of I dopant can reduce the band gap of TiO_2 photoanode and improve the of NKX-2554 dye, which can improve the visible-light absorption of anatase TiO_2 and can also facilitate the electron injection from the dye molecule to the TiO_2 substrate. As a result, the I doping can significantly enhance the incident photon-to-current conversion efficiency (IPCE) of DSSCs.

1 Introduction

Dye-sensitized solar cells (DSSCs) have been attracting much attention because of their high energy conversion efficiency and as a low-cost alternative to commercial solar cells based on silicon.[1,2] In the past decades, great efforts have been made to further improve the energy conversion efficiency in order to commercialize DSSCs successfully. One of the key elements in DSSCs is the semiconductor photoanode, which composed of anatase TiO_2 nanocrystals. The TiO_2 photoanode transfers photoexcited electrons from the dye molecules to the transparent conducting substrate and concurrently allows the electrolytes to diffuse to the anchored dyes.[3] However, such pure TiO_2 nanocrystalline films usually show poor light harvesting due to its wide band gap (3.2 eV for anatase TiO_2). It is well known that enhancing the photocapture efficiency and optical absorption of the photoelectrode films can significantly improve the photon-to-current conversion efficiency of TiO_2-based DSSCs.[4]

Recently, it is reported that the incident photon-to-current conversion efficiency (IPCE) and the overall conversion efficiency of DSSCs can be enhanced by using a non-metal (such as N and I) doped TiO_2 nanocrystalline as photoanode.[4-6] For example, Hou et.al observed that the photovoltaic performance of the DSSCs based on I-doped TiO_2 photoanode is 42.9% higher than that of the cell based on undoped TiO_2 photoanode.[6] The improved performance is explained by the expanded visible-light harvesting, lowered recombination resistance together with prolonged electron lifetime of I-doped TiO_2 photoanode. However, the fundamental physical processes of the IPCE enhancement in

[a] Corresponding author : niumang@upc.edu.cn

the DSSCs based on the I-doped TiO_2 photoanode is not understood completely yet, because the adsorption property of the dye sensitizer on I-doped TiO_2 photoanode and the electronic property of such adsorbed complexes have not been investigated fully.

It is well known that density functional theory (DFT) calculations have been widely used to investigate the adsorption and electronic properties of dye molecule on TiO_2 substrate. Therefore, we performing DFT calculations to discuss the origin of IPCE enhancement in the DSSCs based on the I-doped TiO_2 photoanode. [7] Firstly, we calcutled the electronic structures of pure and I-doped TiO_2(101) surfaces. Then, the absorption behavior of dye molecule on the pure and I-doped TiO_2(101) surfaces and the electronic properties of adsorbed dye/TiO_2 complexes were investigated. Finally, the IPCE enhancement in the DSSCs based on the I-doped TiO_2 photoanode is discussed.

2 Calculation Methods

Our calculations calculations were carried out using the Vienna Ab-initio Simulation Package (VASP)[8] with the generalized gradient approximation (GGA). The first principles pseudopotential were generated by the projector-augment-wave (PAW)[9] method. The Perdew-Burke-Ernzerhof (PBE)[10] of was adopted to describe the exchange and correlation functionals. The cutoff energy for the plane-wave basis set was set to 500 eV. A $3\times3\times1$ Monkhorst-pack grid was used in the DFT calculations. For the geometry optimizations, the convergence criterion of forces on each ion was set to 0.01 eV/Å. The resulting structures were then used for the electronic structures calculations. Considering that the standard DFT often fails to predict the correct band gaps of transition metal oxides like TiO_2 due to the self-interaction error inherent in such functional, we adopted DFT+U method for geometry optimization and subsequent electronic structure calculations. The onsite U = 6.0 eV and J = 0.5 eV parameters were applied to the Ti-3d electrons.[11]

The adsorption of dye molecular on TiO_2 surface was simulated by using vacuum slab models. The TiO_2 surfaces were modeled by a 2×4 anatase TiO_2(101) supercell along [101] and [010] directions with double-layer of slab. We fixed the bottom layer of slab during the geometry optimizations. A 25 Å vacuum region was used to ensure the decoupling between neighboring systems. The I-doped TiO_2(101) slab was constructed by inducing I atom to the Ti site of TiO_2. Such vacuum slab models have been used often in the theoretical calculations of DSSCs.[12] The polyene-dye of 2-Cyano-5,5-bis(4-dimethylaminophenyl)penta- 2,4-dienoic acid (NKX-2554) was used as the photosensitizer in our simulation models because such organic dye has shown high IPCE performance in DSSCs. The adsorption energy of NKX-2554 dye on TiO_2(101) slab, E_{ads}, is calculated using the expression

$$E_{ads} = E_{slab} + E_{dye} - E_{dye/slab} \quad (1)$$

where E_{slab} is the energy of the clean TiO_2(101) slab, E_{dye} represents the energy of the NKX-2554 dye, and $E_{dye/slab}$ represents the total energy of the adsorbed dye/TiO_2 system. A positive value of $E_{slab} > 0$ indicates stable adsorption.

3 Results and Discussion

3.1 I-doped anatase TiO_2(101) surface

To explore the origin of IPCE enhancement in the DSSCs based on the I-doped TiO_2 photoanode, we first investigated how the incorporation of I-dopant affects the electronic structures of the anatase TiO_2. The DFT+U calculated density of states (DOSs) of pure and I-doped anatase TiO_2(101) surface were displayed in Fig. 1a and Fig1b, respectively. The highest occupied state is chosen as the Fermi energy and is displayed with a black dashed line. For the pure anatase TiO_2(101) surface, its valence band (VB) and conduction band (CB) were separated by a wide gap. The valence band consists

DOI: 10.1051/matecconf/20178803002

mainly of O-2p states, including some hybridization with the Ti-3d orbital. The conduction band is dominated by Ti-3d states. Fig. 1b shows that the VB and CB of I-doped anatase $TiO_2(101)$ surface are the same as those of undoped TiO_2. However, I-doping introduces a fully occupied band gap state 0.5 eV above the valence band maximum (VBM). The partial DOS plots shows that this band gap state consists mostly of I 5s states (not shown here) and the 2p-states of its neighboring O atoms, and that the I-5p states contribute to the conduction band. This indicates that I dopant on Ti site exists as an I^{5+} (s2) cation, as found experimentally.[13] The calculated band gap energy is 2.62 and 1.73 eV for the pure and I-doped anatase $TiO_2(101)$ surface, respectively. Therefore, the effective band gap reduction of anatase TiO_2 photoanode through I doping can lead to a significant red-shift of optical absorption edge, which provides a natural explanation for the increased IPCE of DSSCs.

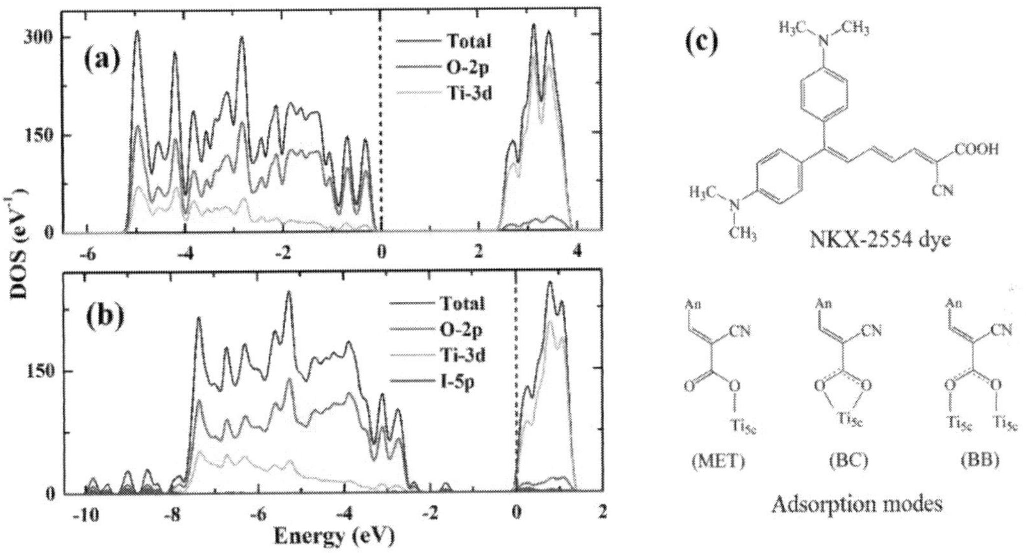

Figure 1. The DFT+U calculated DOSs for (a) pure and (b) I-doped anatase slabs. (c) The possible dissociative adsorption configurations of NKX-2554 dye on $TiO_2(101)$ surface.

In addition, the Fermi level of I doped $TiO_2(101)$ surface (Fig. 1b) is pinned at 0.16 eV above the conduction band minimum (CBM) so that iodine is effectively acting as an n-type dopant in TiO_2, showing a typical n-type metallic characteristic. Therefore, it is suggested that the I-doping can improve the conductivity of anatase TiO_2 photoanode. In fact, the open-circuit voltage (Voc) of DSSCs is determined by the energy difference between the CBM of TiO_2 electrode and the reduction level of hole-transport material (HTM). The n-type I-doping lead to the increase of the electron concentration and upward shift of the CBM, as a result, the Voc of of DSSCs is enlarged. These results are consistent with the corresponding experimental findings.[6]

3.2 Adsorption of NKX-2554 dye on $TiO_2(101)$ surface

In addition to the high solar-cell performance of NKX-2554 dye, its prospective low-cost production owing to its simple molecular structure and synthesis procedure suggest a possibility of promising applications of organic-dye photosensitizers in DSSCs. The NKX-2554 dye is bestowed with a donor part (electron donating group), an acceptor part (electron accepting group) and a anchoring group. The cyanoacrylic acid anchoring group in NKX-2554 has a strong binding to the TiO_2 surface.[14] The adsorption of NKX-2554 dye on anatase (101) surface has been extensively studied to search for the energetically most favorable structure of adsorbed dye/TiO_2 systems.[15] The possible dissociative adsorption modes for the cyanoacrylic acid derivative are shown in Fig. 1c. The coordination may be monodentate ester-type (MET), bidentate chelating (BC), or bidentate briding (BB) depending on the number of oxygens used by the carboxylate anion (An-COO-) to coordinate the five-coordinated surface Ti (Ti5c) acidic sites.

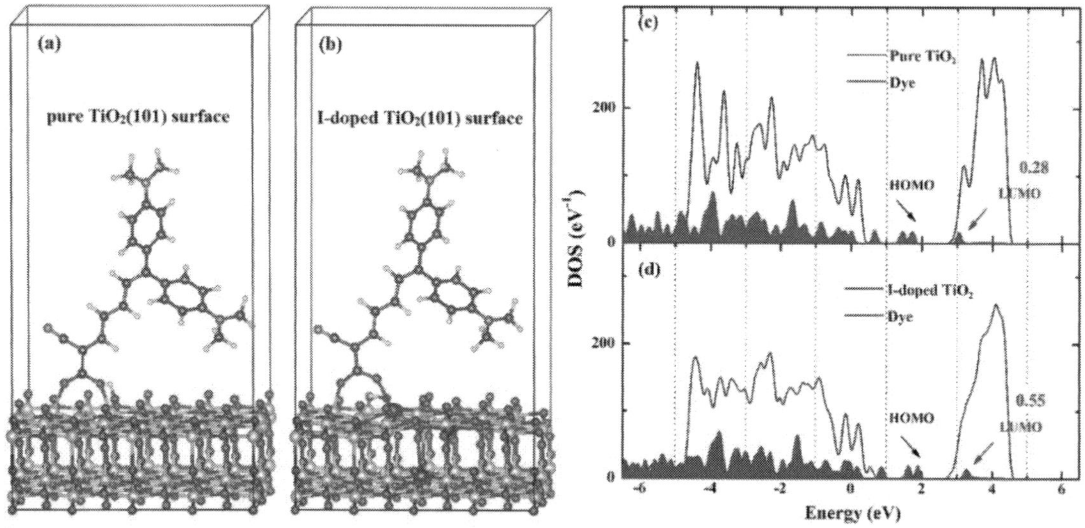

Figure 2. The optimized stable configuration of NKX-2554 dye adsorbed on (a) pure and (b) I-doped anatase $TiO_2(101)$ slabs. The DFT+U calculated DOSs for (c) adsorbed NKX-2554 dye/pure $TiO_2(101)$ slab and (d) adsorbed NKX-2554 dye/I-doped $TiO_2(101)$ slab.

The calculated Eads energies of these adsorption configurations indicate that the BB-type configuration is the most favorable mode of dissociative adsorption for NKX-2554 dye,[15] with a positive Eads of 1.29 eV. The optimized geometrical structure of the NKX-2554 dye adsorbed on pure and I-doped $TiO_2(101)$ slab are given in Fig. 2a and Fig. 2b, respectively. It is found that although the incorporation of I-dopants in $TiO_2(101)$ slab lead to a slight deformation of the surface structure, the stable adsorption configuration of the NKX-2554 dye is not changed compared with that of pure $TiO_2(101)$ surface.

The DFT+U calculated DOSs for the adsorbed NKX-2554 dye/$TiO_2(101)$ surface (or I-doped $TiO_2(101)$ surface) complexes are presented in Fig. 2c and Fig. 2d, respectively. As shown in Fig. 2c, the NKX-2554 dye induces sharp occupied molecular energy levels in the band gap of $TiO_2(101)$ slab after adsorption. The highest occupied molecular orbital (HOMO) of the dye is uncoupled with the valence band of the TiO_2 surface. However, the lowest unoccupied molecular orbital (LUMO) of the dye is coupled with the valence band of the TiO_2 surface. In adsorbed dye/TiO_2 complexes, the LUMO locates 0.28 eV above the CBM level of the $TiO_2(101)$ slab, which could facilitate a fast electron injection from the dye LUMO to the conduction band of TiO_2.[16] Fig. 2d shows the DOSs plots of NKX-2554 dye adsorbed on I-doped $TiO_2(101)$ slab. Similarly, the LUMO of the dye is coupled with the valence band of the I-doped $TiO_2(101)$ surface. It is found that the dye LUMO is 0.55 eV above the CBM of I-doped $TiO_2(101)$ slab. Therefore, the I-doping in TiO_2 photoanode lead to the increasement of dye LUMO position, which can improve the electron injection rate and the performance of DSSCs.

4 Summary

In summary, we have used DFT+U calculations to explore the mechanism of IPCE enhancement in DSSCs based on I-doped anatase TiO_2 photoanodes. It is indicated that the I-doped anatase $TiO_2(101)$ surface show n-type metallic characteristic, which can improve the conductivity of anatase TiO_2. The uphill shift of CMB position in I-doped anatase $TiO_2(101)$ surface can increase the open-circuit voltage of DSSCs. In addition, the band gap of $TiO_2(101)$ surface is effectively reduced through I-doping, and thus enhance the visible-light absorption of TiO_2 photoanode. Moreover, It is also found that the I-doping in anatase $TiO_2(101)$ surface can facilitate the electron injection from the NKX-2554 dye to the TiO_2 substrate by analyzing the calculated electronic properties of adsorbed dye/TiO_2 complexes. Therefore, it is concluded that I doping can significantly enhance the IPCE of DSSCs due to the improved conductivity, increased open-circuit voltage, extended photo-response range of TiO_2 photoannode, and the effective electron injection of adsorbed dye/TiO_2 complexes.

Acknowledgements

This work is supported by the Fundamental Research Funds for the Central Universities.

References

1. B. O'regan, M. Grätzel, Nature **353**, 737(1991)
2. M. Grätzel, Nature **414**, 338(2001) 338-344.
3. T. Horiuchi, H. Miura, K. Sumioka, S. Uchida, J. Am. Chem.l Soc. **126**, 12218(2004)
4. T. Ma, M. Akiyama, E. Abe, I. Imai, Nano Lett. **5**, 2543(2005)
5. M. Samadpour, P.P. Boix, S. Giménez, A. Iraji Zad, N. Taghavinia, I. Mora-Seró, J. Bisquert, J. Phys. Chem. C **115**, 14400(2011)
6. Q. Hou, Y. Zheng, J.F. Chen, W. Zhou, J. Deng, X. Tao, J. Mater. Chem. **21**, 387(2011)
7. M. Niu, R. Cui, H. Wu, D.J. Cheng, D.P. Cao. J. Phys. Chem. C **119**, 13425(2015)
8. G. Kresse, J. Furthmüller, Phys. Rev. B **54,** 11169(1996)
9. P.E. Blöchl, Phys. Rev. B **50**, 17953(1994)
10. B. Hammer, L.B. Hansen, J.K. Nørskov, Phys. Rev. B **59**, 7413(1999)
11. C.J. Calzado, N.C. Hernández, J.F. Sanz, Phys. Rev. B **77**, 045118(2008)
12. F. De Angelis, S. Fantacci, A. Selloni, M.K. Nazeeruddin, M. Grätzel, J. Phys. Chem. C **114,** 6054(2010)
13. X. Hong, Z. Wang, W. Cai, F. Lu, J. Zhang, Y. Yang, N. Ma, Y. Liu, Chem. Mater. **17**, 1548(2005)
14. K. Hara, M. Kurashige, S. Ito, A. Shinpo, S. Suga, K. Sayama, H. Arakawa, Chem. Commun., 252(2003)
15. K. Srinivas, K. Yesudas, K. Bhanuprakash, L. Giribabu, J. Phys. Chem. C **113,** 20117(2009)
16. T. Bessho, S. M. Zakeeruddin, C. Y. Yeh, E. W. G. Diau, M. Grätzel, Angew. Chem. **122**, 6796(2010)

Electronic and Elastic Properties of $CaMg_2$ Alloy Phase under Various Pressures by Density Functional Theory

Jia Fu[1,a], Weihui Lin[2], Zhongren Chen[1] and Keqiang Wang[1]

[1] *Department of Material Chemistry, Southern University of Science and Technology, Shenzhen 518055, China*
[2] *School of Mechanics College, Taiyuan University of Technology, Taiyuan 030024, China*

Abstract. The influencing effect of pressure on structural, elastic and electronic properties of $CaMg_2$ Laves phase is mainly investigated. The optimized structural parameters at zero pressure are a=b=6.250Å, c=10.101Å, which has good agreement with the experimental and theoretical values. The elastic constants are calculated, and then the bulk modulus, shear modulus, Young's modulus, Poisson's ratio and anisotropy factor are determined. The results show that the applied pressure is beneficial to the elastic properties of $CaMg_2$. The analysis of electronic density of states (DOS) and Mulliken electron population reveal the bonding characteristics in $CaMg_2$ crystal. Finally, the Debye temperatures under different pressures are obtained from the average sound velocity.

1 Introduction

Magnesium alloys as low density, high specific strength, good stiffness and strength-weight ratio are widely used in aerospace industry, automobile industry, 3D printing and other fields[1,2]. Magnesium alloys can be significantly improved by adding rare-earth (RE) elements[3] and Ca element[4]. Mg-Ca alloy exhibits excellent oxidation- resistance and burning-resistance[5], and it is also regarded as one of the most promising alloy due to its appreciable corrosion resistance[6]. $CaMg_2$ Laves phase as its excellent physical and chemical properties is a typical phase in Mg-Y-Ni[7] and Mg-Al-Ca alloy[8]. Yu et al. [9] investigated the electronic structure and elastic properties of the binary AB2 type Laves phase in Mg-Al-Ca alloys. Elastic constants of $CaMg_2$ alloy are studied by Tang et al.[2]. Zhou et al. [6] has investigated the phase stability and electronic structure of an Mg-Ca system, showing that $CaMg_2$ Laves phase is ductile with a structural stability. However, there is no report about the experimental values of Debye temperature and the effect of pressure on structural, elastic and electronic properties of $CaMg_2$ Laves phase. In this work, the first-principles calculations are used to investigate the anisotropic elasticity and electronic properties of the $CaMg_2$ Laves phase under pressure. Moreover, the density of state (DOS) under different pressure is analysed to insight into the bonding characteristics of $CaMg_2$, and then reveals the underlying structural stability mechanism of $CaMg_2$ Laves phase. The current work can act as a theoretical prediction and provides mechanical parameters of $CaMg_2$ alloy.

2 Theoretical principle and calculation algorithm

[a] Corresponding author : fuj@sustc.edu.cn

2.1 Description of elastic constants algorithm and Young's modulus calculation

A multi-particle electronic structure satisfies the Schrödinger equation. Generally, Hamiltonian contains kinetic energy, Coulomb potential energy between electrons and the external potential, which is described as:

$$H=\sum_{i}-\frac{h^{2}}{2m}\nabla_{i}^{2}+\sum_{i,j}{}'\frac{e_{1}^{2}}{\left|\vec{r}_{i}-\vec{r}_{j}\right|}+\sum_{i}V_{i}\left(\vec{r}\right) \qquad (1)$$

Where, $V_i(r)$ is an external potential; r_i and r_j are the nucleus position vectors; while mε and m stand respectively for the mass of nuclei and electrons. h is the reduced Planck constant.

Kohn-Sham equation as an approximation to simplify Schrödinger equation is described as[10]:

$$\begin{cases}\left[-\frac{h}{2m}\nabla^{2}+V_{eff}\left(\vec{r}\right)\right]\varphi_{i}\left(\vec{r}\right)=\varepsilon_{i}\varphi_{i}\left(\vec{r}\right)\\ V_{eff}\left(r\right)=V\left(\vec{r}\right)+V_{ee}\left(\vec{r}\right)+V_{ex}\left(\vec{r}\right)\\ V_{xe}\left(\vec{r}\right)=\frac{\delta E_{xe}\left[\rho\right]}{\delta\left(\vec{r}\right)}\\ \rho\left(\vec{r}\right)=\sum_{i}\left|\phi_{i}\left(\vec{r}\right)\right|^{2}\end{cases} \qquad (2)$$

Where, $V_{xe}\left(\vec{r}\right)$ is an exchange-correlation potential; $\rho\left(\vec{r}\right)$ is the electron density. Equation (2) includes not only the nucleus and electron kinetic energy term but also term of nucleus- nucleus and electron-electron interactions[10]. The electron density is optimised and elastic constants are obtained.

For hexagonal structure, mechanical stability condition and Voigt/ Reuss bounds of B and G are[11]:

$$C_{11}>0;C_{11}-C_{12}>0;C_{44}>0;(C_{11}+C_{12})C_{33}-2C_{13}^{2}>0 \qquad (3)$$

$$B_{V}=\frac{2}{9}(C_{11}+C_{12}+\frac{1}{2}C_{33}+2C_{13});\ G_{V}=\frac{1}{30}(7C_{11}-5C_{12}+12C_{44}+2C_{33}-4C_{13}) \qquad (4)$$

$$B_{R}=\frac{(C_{11}+C_{12})-2C_{13}^{2}}{C_{11}+C_{12}+2C_{33}-4C_{13}};\qquad G_{R}=\frac{5}{2}\left\{\frac{[(C_{11}+C_{12})C_{33}-2C_{13}^{2}]C_{44}C_{66}}{3B_{V}C_{44}C_{66}+[(C_{11}+C_{12})C_{33}-2C_{13}^{2}](C_{44}+C_{66})}\right\} \qquad (5)$$

Then homogenized elastic properties of polycrystals can be derived. Shear and bulk moduli are obtained by means of Voigt and Reuss bounds, which are based on the C_{ij} calculated. Then the Voigt-Reuss-Hill average and Young's modulus can be determined from shear/bulk modulus[11].

Based on C_{ij}, Debye temperatures can be estimated by the some sound velocity equations[12]:

$$\Theta_{D}=\frac{h}{k_{B}}\left[\frac{3n}{4\pi}\left(\frac{N_{A}\rho}{M}\right)\right]^{1/3}V_{m},\ V_{m}=\left[\frac{1}{3}\left(\frac{2}{V_{s}^{3}}+\frac{1}{V_{l}^{3}}\right)\right]^{-1/3},\ V_{s}=\sqrt{\frac{G}{\rho}},\ V_{l}=\sqrt{\left(B+\frac{4}{3}G\right)\frac{1}{\rho}} \qquad (6)$$

Where Θ_D is Debye temperature; h and k_B are the Planck and Boltzmann constant; n the total number of atoms; ρ is density; N_A the Avogadro number; M the molecular weight per formula; V_m the average sound velocity; V_s the transverse sound velocity and V_l the longitudinal sound velocity.

2.2 Desription of CaMg2 modeling and computational method

Fig.1 shows $CaMg_2$ crystal of hexagonal structure with a space group P63/mmc. Ca atoms are located on *4f* sites and Mg atoms *2a*, *6f* sites. Pseudo atomic calculations are performed for Mg *2p6 3s2* and Ca *3s2 3p6 4s2*.

Crystal lattice (a=b=6.250Å, c=10.101Å, α=β=90°, γ=120°) is shown in Fig.1 a). DFT calculation is shown in Fig.1 b), initial conditions are as: the exchange-correlation function is generalized gradient approximation (GGA); Cutoff energy of the plane waves for ultrasoft pseudopotentials is 400 *eV*.

Brillouin zone is 12×12×8 by Monkhorst-Pack method. The self-consistent convergence of the total energy is $5.0\times10^{-6}eV$ per atom and convergence of the force on all atoms is less than 0.001eV/nm, the pressure region is 0～$60GPa$.

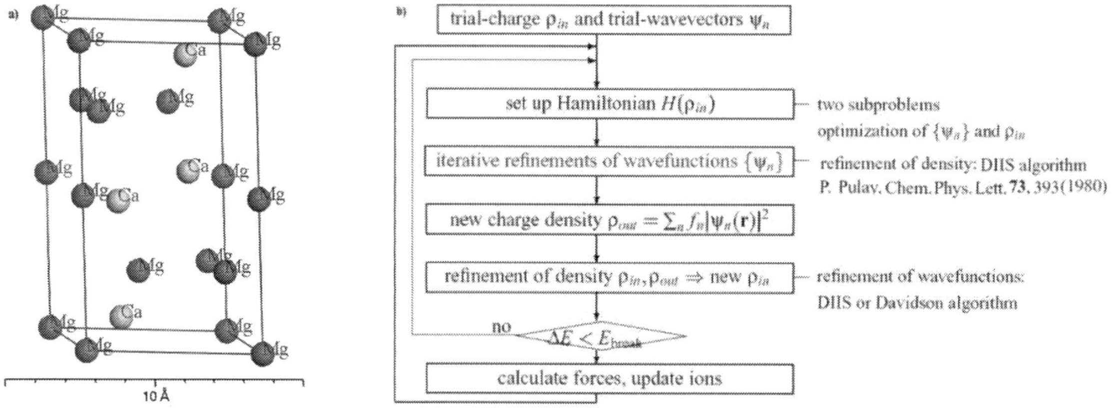

Figure.1 a) Final parameters of $CaMg_2$ crystal; b) Flowchart of DFT by solving Kohn-Sham equations

3 Results and discussion

3.1 Structural parameters and elastic constants

The equilibrium lattice parameters, elastic constants C_{ij} and bulk modulus B_0 are in Tab.1.

Table.1 The structural parameters (a_0, c_0, c_0/a_0, V_0), elastic constants and bulk modulus B_0 of $CaMg_2$ phase at 0GPa.

a) Lattice parameters

	a_0(Å)	c_0(Å)	c_0/a_0	V_0(Å^3)
This work	6.250	10.101	1.616	341.732
Cal.[10]	6.240	10.140	1.625	342.200
Cal.[2]	6.232	10.093	1.620	339.740
Exp.[7]	6.220	10.100	1.624	341.081

b) Elastic constants

	C_{11}	C_{12},	C_{13}	C_{33}	C_{44}
This work	55.70	18.05	15.26	58.39	16.95
Cal.[9]	51.43	22.31	14.73	58.51	14.32
Cal.[3]	62.95	15.27	13.64	65.20	17.77

As the values in Tab.1 match well with experimental values, it is indicated that the present model is reliable. Variations of lattice parameter and elastic constants with increase of pressure are in Fig.2.

DOI: 10.1051/matecconf/20178803003

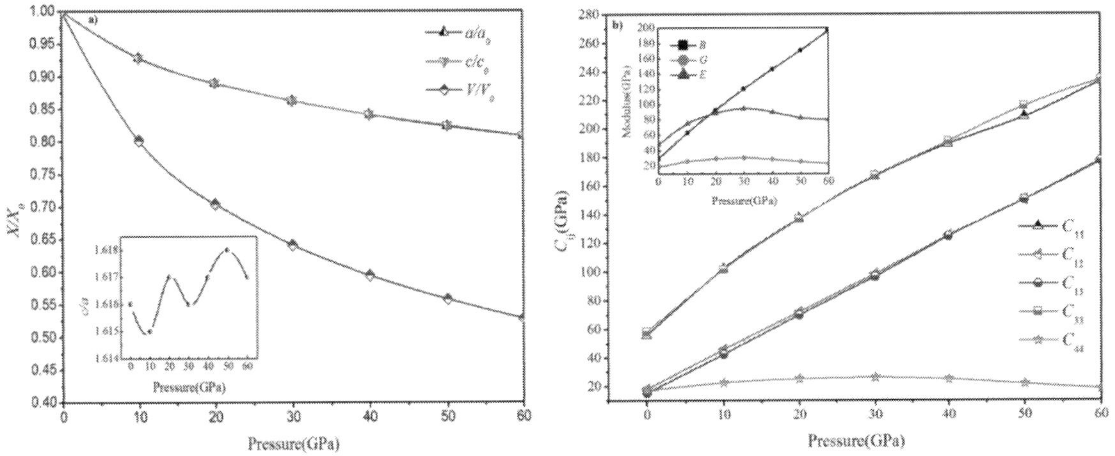

Figure.2 Parametres under pressure of 0- 60GPa: a) Relative change of a, c and V; b) Elastic constants C_{ij}.

As shown in Fig.2 a), it can be seen that $CaMg_2$ has not undergone a phase transition according to the smooth variation of a/a_0, c/c_0 and V/V_0 with the increase of pressure. Moreover, it shows that the lattice parameter a and parameter c are almost equal throughout the whole pressure range, indicating that the elastic anisotropy of $CaMg_2$ is weak. The compressive ability of $CaMg_2$ crystal under pressure is consistent with the variation trend of the V/V_0 curve. The axial ratio c/a changes little with increase of pressure, from minimum 1.615 to maximum 1.618. As in Fig. 2 b), we can conclude that C_{11} and C_{33} are more sensitive to the change of applied pressure than C_{12} and C_{13}. Besides, the applied pressure has less effect on C_{44} and G, while C_{11}, C_{12}, C_{13}, C_{33}, bulk and Young's modulus show the increasing tendency with the increase of pressures.

3.2 Prediction of elastic moduli of CaMg2 ploycrystals

To investigate the effect of the applied pressure on structural properties of $CaMg_2$, The changes of Poisson's ratio and G/B with increasing pressure are plotted in Fig.3

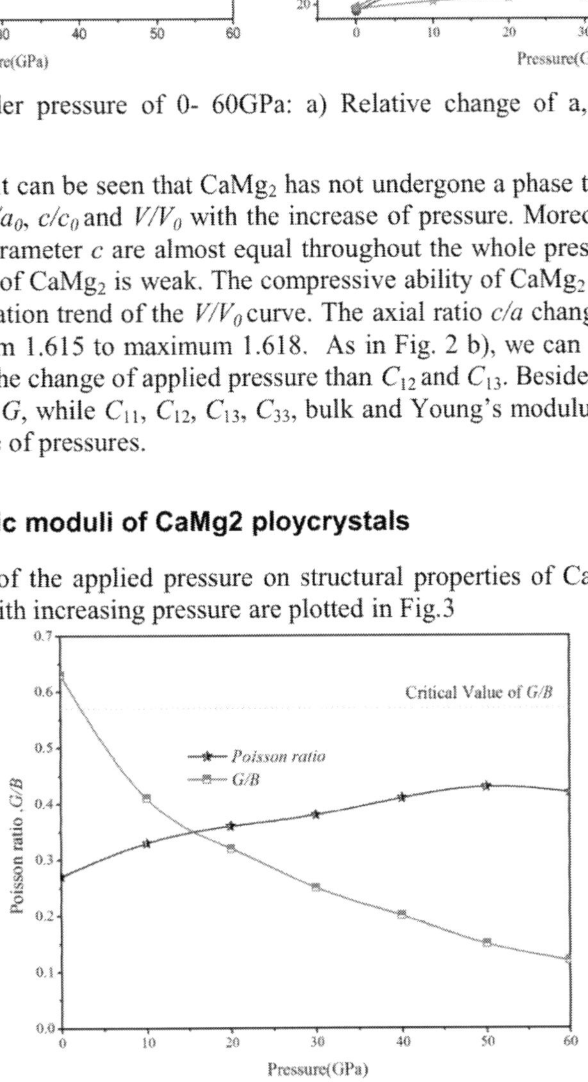

Figure.3 The pressure independence of Poisson's ratio and the G/B ratio of $CaMg_2$

As shown in Fig.3, the G/B ratio decreases with the increasing pressure. Poisson's ratio at $0GPa$ is 0.27, showing CaMg2 is with predominantly central interatomic forces. The critical value which separates brittleness and ductility is about 0.57 in the horizontal dotted line. Evidently, the G/B ratio within region of $0\sim4GPa$ is higher than critical value 0.57 at $4GPa$, which implies that $CaMg_2$ phase show a little brittle at $0GPa$ (G/B =0.63) and turn to be a ductile behavior after critical value. It is indicated that $CaMg_2$ shows slight brittleness under 0～4*GPa* and good ductility under 4～60*GPa*.

3.3 Analysis of the density of state (DOS) and bonding characteristics

Fig.4 shows the total and partial DOS of $CaMg_2$ phase near Fermi level under 0*GPa* in Fig.4 a), 20*GPa* in Fig.4 b), 40*GPa* in Fig.4 c) and 60*GPa* in Fig.4 d).

As in Fig.4, the total DOS of $CaMg_2$ is mainly made up of the *s, p* states of Mg atoms below Fermi level and the *s, p* states of Ca atoms above Fermi level. It is evident that the total DOS of $CaMg_2$ decreases with elevated pressure, which implies that applied pressure is of benefit to the stability of $CaMg_2$. Moreover, $CaMg_2$ is characterised to have the metallic behaviour as there is no energy gap near the Fermi level[2,6,8,9]. The main bonding peak is located in the energy region from -2 to 4eV, originated from the contribution of valence electron numbers of Mg *s*, Mg *p*, Ca *s* and Ca *p* orbits. A hybridization of Mg *s, p* and Ca *s, p* orbits is seen in Fig.4 a), indicating the existence of covalent bonding in $CaMg_2$ crystal. From Fig.4 b) to Fig.4 d), the shifting of Mg *s* and *p* orbits from higher to lower energy region is discovered. Moreover, with increasing pressure, the Ca *s* and *p* orbits present a slight broadening and a decrease of intensity, which indicates that the hybridization of Mg *s, p* and Ca *s, p* orbits is becoming stronger. It is worthwhile to note that the applied pressure has a little effect on the DOS when the pressure is above 40*GPa*.

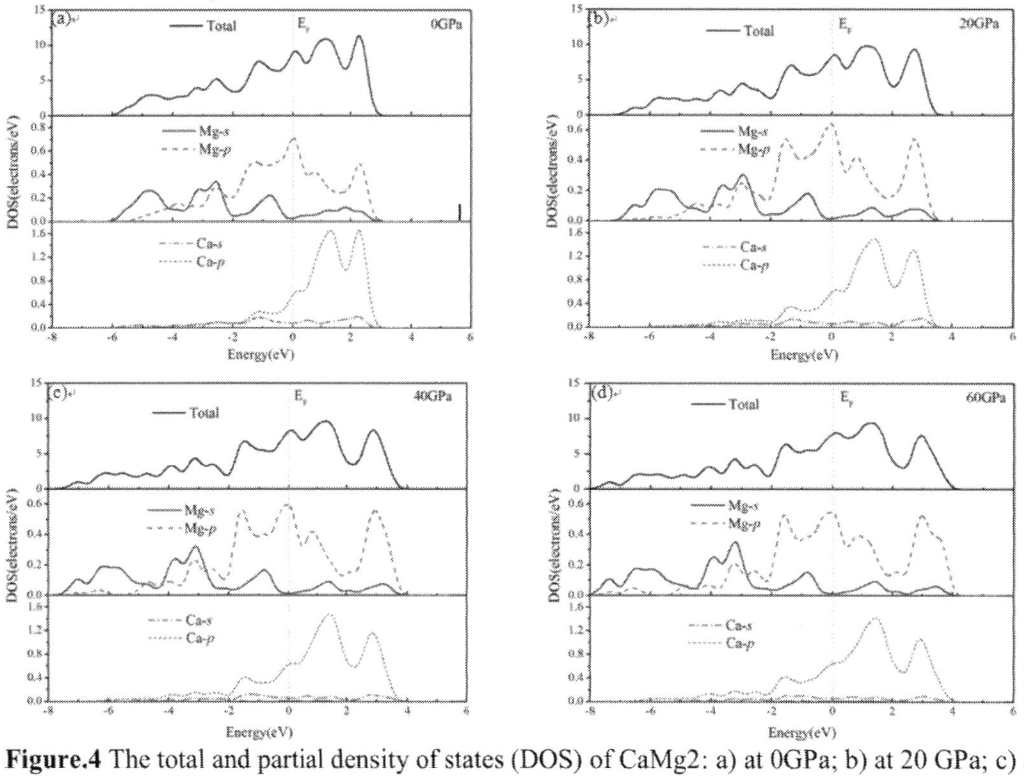

Figure.4 The total and partial density of states (DOS) of CaMg2: a) at 0GPa; b) at 20 GPa; c) at 40 GPa; d) at 60 GPa.

3.4 Analysis of Mulliken electron population and Debye temperature

The Mulliken electron population has been performed to consider the ionic bond feature of $CaMg_2$ phase. Table 4 shows the Mulliken electron population of $CaMg_2$ under pressure.

 DOI: 10.1051/matecconf/20178803003

Table 2: a) Mulliken electron population of pressure; b) Pressure dependence of parameter ρ, V_s, V_l, V_m and Θ_D

Pressure(GPa)	Species	s	p	d	Total	Charge (e)
	Ca	2.44	6.00	0.66	9.10	0.90
0	Mg (I)	0.92	7.45	0.00	8.37	-0.37
	Mg (II)	0.93	7.55	0.00	8.48	-0.48
	Ca	2.32	5.98	1.00	9.30	0.70
30	Mg (I)	0.89	7.31	0.00	8.20	-0.20
	Mg (II)	0.90	7.50	0.00	8.40	-0.40
	Ca	2.27	5.96	1.25	9.48	0.52
60	Mg (I)	0.93	7.09	0.00	8.02	-0.02
	Mg (II)	0.95	7.39	0.00	8.34	-0.34

Pressure (GPa)	ρ (kg/m^3)	V_s (m/s)	V_l (m/s)	V_m (m/s)	Θ_D (K)
0	1.724×10^3	3277.83	5614.24	3635.41	348.70
10	2.150×10^3	3470.14	6249.48	3865.02	399.02
20	2.447×10^3	3462.69	7288.72	3895.40	419.86
30	2.688×10^3	3372.35	7744.53	3808.73	423.56
40	2.898×10^3	3151.89	7987.80	3571.81	407.29
50	3.086×10^3	2886.38	8146.42	3279.92	381.92
60	3.260×10^3	2656.17	8420.52	3024.84	358.71

(Note: ρ is density, V_s transverse sound velocity, V_l longitudinal sound velocity, V_m average sound velocity, Θ_D Debye temperature.)

As in Tab.2 a), the total charge transfer from Ca atoms to Mg atoms at 0*GPa* is 3.60, at 30*GPa* is 2.80 and at 60*GPa* is 2.08. The charge transfer from Ca atoms to Mg atoms demonstrates the ionic bonding in $CaMg_2$ crystal. It is found that the charge transfer from Ca to Mg decreases with the increasing pressure, implying that the ionic bonding is getting weaker[8] under the higher pressure.

Tab.2 b) has listed the calculated V_s, V_l, V_m and Θ_D from 0～60 GPa of $CaMg_2$. Debye temperature at 0 GPa is 348.70 K, and it is clear that the Debye temperature increases under 0～30*GPa* with increasingly applied pressure then presents a declining trend when the applied pressure is above 30*GPa*.

4 Conclusions

Structural, elastic and electronic properties of $CaMg_2$ Laves phase under external pressure from 0～60*GPa* are investigated by DFT. Conclusions are as follows:

1) The optimized structural parameters (a=b=6.250Å, c=10.101Å, V=341.732 Å^3) are very close to experimental values. From DFT calculation, $CaMg_2$ Laves phase under 0～60GPa does not undergo a phase transition, and the compressibility in *a* direction is better than in *c* direction.

2) Based on elastic constants by DFT, the mechanical parameters of $CaMg_2$ are as: bulk modulus is 29.659*GPa*, shear modulus 18.553*GPa*, Young's modulus 49.286*GPa*, Poisson's ratio 0.196 and elastic anisotropy factor *A*. The *G*/*B* ratio shows a decreasing tendency with the increase of pressure.

3) From DOS analysis at 0, 20, 40 and 60*GPa* show that below Fermi level the total DOS is mainly Mg *s*, *p* orbits while above Fermi level the total DOS is almost Ca *s*, *p* orbits. Moreover, covalent bonding is existed in CaMg2 crystal according to the orbital hybridization of Mg *s*, *p* and Ca *s*, *p* orbits. Mulliken electronic population demonstrate the charge transfer from Ca to Mg becomes less thus the ionic bond between Ca and Mg becomes weaker with the increase of pressure.

4) The Debye temperature shows an increasing tendency with the applied pressure from 0 ～30

GPa and then presents a declining tendency under the pressure region of 30 GPa～60GPa.

References

1. Z.B. Sajuri, T. Umehara,Y. Miyashita et al., Advanced Engineering Materials, **5** (2003) 910-916.
2. B.Y. Tang, W.Y.Yu, X.Q. Zeng et al., Materials Science and Engineering: A, **489** (2008) 444.
3. B.L. Mordike, T. Ebert, Materials Science and Engineering: A, **302** (2001) 37–45.
4. K.Oh-Ishi, C.L.Mendis, T.Homma, S.Kamado et al., Acta Materialia, **57** (2009)5593-5604.
5. J.F. Fan, G.C. Yang et al., Chinese Journal of Nonferrous Metals, 2004, **14** (10):1666-1670.
6. P. Zhou et al., Journal of the Mechanical Behavior of Biomedical Materials, **8** (2012) 154-164.
7. K. Aonol, S. Orimo, H. Fujii, Journal of Alloys and Compounds., **309** (1-2), (2000),L1-L4.
8. A. Suzuki, N.D. Saddock, J.W. Jones, T.M. Pollock, Acta Materialia, **53** (2005) 2823-2834.
9. W.Y. Yu, N.Wang, X.B. Xiao, B.Y. Tang et al., Solid State Sciences, **11** (2009):1400.
10. Aaeid A., PhD Thesis, l'Université des Sciences et Technologies de Lille, 2010 France. (French)
11. Wu Z J et al., Physical Review B, **76** (2007), 054115
12. O.L.Anderson, Journal of Physics and Chemistry of Solids, **24** (1963) 909.

AUTHOR INDEX